Earth, air, fire, and
water are obedient creatures,
they are dead to you and me,
but alive at God's presence.

—Rumi

CERAMIC HOUSES

How to Build Your Own

NADER KHALILI

1817

Harper & Row, Publishers, San Francisco

Cambridge, Hagerstown, New York, Philadelphia, Washington
London, Mexico City, São Paulo, Singapore, Sydney

Persian language poem calligraphy by F. Haydari.

"Shell Membrane Theory Applied to Masonry Domes" by Professor Zareh B. Gregorian has appeared in *Art & Architecture* (Iran) and is reprinted with permission. "Magma, Ceramic, and Fused Adobe Structures Generated *In Situ*" was originally delivered at a NASA symposium "Lunar Base and Space Activities of the 21st Century."

FIRST EDITION

Designer: Donna Davis

Library of Congress Cataloging-in-Publication Data

Khalili, Nader.
Ceramic houses.

1. Adobe houses—Iran—Bushrūyah. 2. Earth houses—Iran—Bushrūyah. 3. Glazing (Ceramics)—Iran—Bushrūyah. 4. Vernacular architecture—Iran—Bushrūyah. 5. Buildings—Iran—Bushrūyah—Earthquake effects. Bushrūyah (Iran)—Buildings, structures, etc. 7. Adobe houses—New Mexico—Taos. 8. Earth houses—New Mexico—Taos. 9. Glazing (Ceramics)—New Mexico—Taos. 10. Vernacular architecture—New Mexico—Taos. 11. Buildings—New Mexico—Taos—Earthquake effects. 12. Taos (N.M.)—Buildings, structures, etc. I. Title.

NA7423.B87K43 1986 728-3'73'0924 84-48224

ISBN 0-06-250447-9
ISBN 0-06-250446-0(pbk)

---

86 87 88 89 90 MPC 10 9 8 7 6 5 4 3 2 1

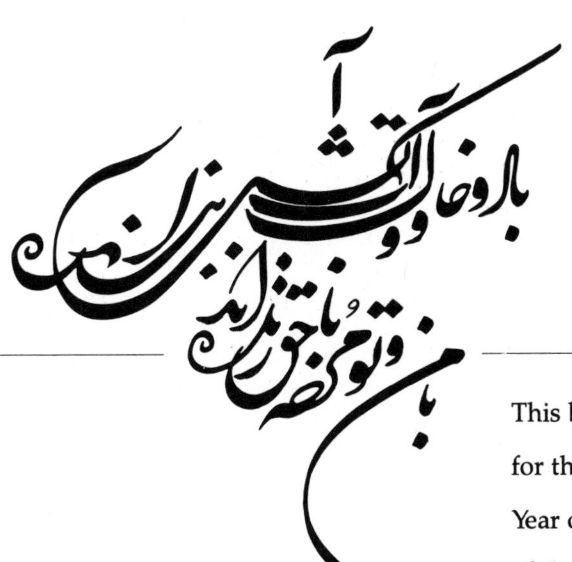

This book is dedicated in the spirit of support for the goal of the United Nations' International Year of Shelter for the Homeless, 1987—that all of the poor and disadvantaged of the world will be able to obtain a home by the year 2000.

CONTENTS

PREFACE

There is an ever-present need for human shelter, especially for the poor. The hopeful movements of the Third World peoples toward understanding the importance of their own identities and indigenous ways; the encouraging words and actions of the Western world—especially the United States—to use the earth, sun, wind, and natural forces; and the rise of general world consciousness to learn these fundamentals; all have helped make me feel a sense of urgency to share my experience, and to teach what I have learned from others.

My main concern is architecture and the people who cannot afford an architect, cannot afford manufactured building materials, cannot afford anything but their own hands and the earth beneath their feet. Anybody in this world should be able to build a shelter for his or her family with the simplest of materials, available to all: the Elements—Earth, Water, Air, and Fire. A family should be able to learn the techniques, move to a piece of empty land, and then—with some water and simple tools—build themselves a house using the earth under their feet. That simple yet profound technology exists today. We have inherited it from our ancestors and must learn how to use and improve on it.

To build a simple house we need not cut trees, weld steel, or buy cement and plastic; in a great many cases the earth alone can suffice. Today the big promises of technology as a cure-all are drifting in rhetorical limbo. Much of the International Style and modern and postmodern architecture is struggling to survive the test of time. There is a new consciousness about the ecology, environment, energy, homelessness, and other basic human issues. To deal with such vital issues effectively we must first understand the fundamentals: the earth and the elements.

In this book I have tried first to talk about the fundamentals. These are covered in the first two parts: "Evolution" details the last ten years of our research with earth and fire and contains unpublished pictures and sections related to my previous book *Racing Alone.* Following is "Philosophy and Design Principles," an exploration of basics of earth architecture. Parts three through five cover the design possibilities, technical aspects, and "how-to" works, such as how to construct buildings using earth as the only material. Although the information in this book is based on the simplest forms of traditionally successful earth architecture, we also explore new possibilities: firing and glazing of the earth structures. Even though firing of kilns—some as large as small houses—has been done in many parts of the world, firing and glazing of buildings for human habitation is a new dimension in earth architecture. The step-by-step making of clay models in Part 5, "Beginning," can put earth structure techniques within the easy reach of young and old everywhere. We can begin to learn at home and school. The final section, "Visions for the Future," shows how far we can take these timeless materials and timeless principles.

The excitement about the earth and fire possibilities has been so immense in my heart that I could not wait to share my experience. Therefore I beg forgiveness for not waiting until we had the necessary details for a perfected system. We still need more experiments and tests, but by sharing information about our works, I hope to draw on the knowledge of the world community, especially in earth architecture and ceramics, to help the idea move towards its perfection.

To create safe and beautiful shelters for large numbers of people, great forces are needed. One of the greatest forces—ready and able—are potters and ceramists. Many regions of the world may not have a painter, weaver, sculptor, jeweler, or even a musician; but there is almost always a potter. The simple village potter who works with raw clay and dung-fire, the sophisticated ceramist who sculpts huge forms and glazes in high-temperature fire, and the millions in between are an untapped force who can simultaneously lead and support the architects and builders, while they realize their own arts in the body of a house. Many of these artists are already making their ceramic pieces as large as small rooms; then why not make the room itself?

> Now the time has come to create a new scale in the
> ceramic world, to walk out from the womb of a pot
> to the space of a room.
>   Now the time has come to step back into history, and
> touch our fathers who have touched the glaze,
> to recapture their secrets of heavenly textures and
> colors, but grow larger than their size,
> to create, not only the little forms and shapes in
> containers, but the spaces that contain us.
>   Now the time has come to create a ceramic glaze, a
> china, a stoneware, not in the scale of our hands
> but in the scale of our lives.*

*Nader Khalili, *Racing Alone* (San Francisco: Harper & Row Publishers, 1983). All quotations in this book not otherwise identified are from *Racing Alone*.

ACKNOWLEDGMENTS

Ten years ago I changed the direction of my life to follow a quest. The tangible results of that decade of search are a few unique structures and two books. The first book, *Racing Along*, recounts the process of my quest, and this one, *Ceramic Houses*, presents some of its products.

The methods and products presented in this book are the result of the work of many, many people. I am grateful for their genuine support and inspiration, and ask forgiveness for not being able to individually include all their names here.

The models and the drawings of this book were prepared by my talented and hard-working student Allessandra Runyon, and the drawings have been brought to their final forms by my colleague Ahde Lahti.

The Geltaftan Team: engineer Manuchehr Sedehi, Mahmoud Hejazi, and Ezzatollah Salmanzadeh (then students of architecture, and now architects), were the permanent members of our group to the end. They worked with utmost dedication and spent hundreds of hours of their time as volunteers in the harshest conditions. Ali Gourang (Ali Aga), the 66-year-old traditional kiln operator and ceramist whose knowledge and sense of humor kept us going, was the first potter in our team—indeed, in the world—to fire and glaze a house. Ostad Asghar, the village mason, undertook the construction of the school, and he taught us his simple ways with earth architecture. Aliakbar Khorramshahi has given great support to our program at the Geltaftan Foundation and has participated in the research and photographics expeditions. Parviz Taidi and Ezzatollah Salmanzadeh participated in the expeditions and provided hundreds of slides and photographs, some of which have been used in this book, but could not be identified separately. Projects presented in the book were photographed by the author; the Boshrouyeh photos were provided by E. Salmanzadeh; and the Kashan photos were provided by P. Taidi.

My students at the Southern California Institute of Architecture (SCI-ARC) have contributed their time and talent to further the research in this work. Some of them have been named in the book in relation to the projects. Native Americans of the Southwest in general and the Institute of American Indian Arts, in Santa Fe, New Mexico, in particular have greatly supported and encouraged our work. Dar-Al-Islam in Abiquiu, New Mexico, has always welcomed me and my students. Some of the photographs used in this book were taken at this complex, which was designed by Hassan Fathy, the Egyptian architect.

The following publications in Persian and English were most helpful in my preparation of this book: *Osul-e-Fanni Sakhteman*, by Professor Mahmoud Maher-Alnaghs; *Honar va Mimari* magazines; articles by Dr. Karim Pirnia, *Asar* publications; *Tarikh-e-Mohandesi dar Iran*, by Dr. Mehdi Farshad; *Sakht-e-Shahr va Mimari dar Eghlim-e-Garm va Khoshk-e-Iran*, by Professor Mahmoud Tavassoli; publications by Freydoun Djoneydi, Bonyade Neychabour; *The Architecture for the Poor*, by Dr. Hassan Fathy; *Sense of Unity*, by Nader Ardalan, architect, and Laleh Bakhtiar; *Adobe, Build It Yourself*, by Paul Graham McHenry, Jr., AIA; *Kilns*, by Professor Daniel Rhodes; articles by Professor Mehdi Bahadori.

I would like to show my gratitude for the spirit of the friendship and support shown for my work and words to SCI-ARC director Ray Kappe, architect Moira Moser, Dr. Paolo Soleri, Professor Peter Blake, publisher Clayton Carlson, architect Hasan-Uddin Khan, architect Charles Correa, Dr. David Warren, Dr. Rena Swentzell, Lorna Bernaldo, M.D., Professor Harry Van Oudenallen, Professor David Stea, architect Arthur Erikson, Harry Kislevitz, James Danisch, Debra Denker, Professor John Russell, Rose Marie Rabin, Ray Meeker, Debrah Smith, Matts Myhrman, and my Native American student and pal Tsosie Tsinhnahjinnie.

My wife, Shiva, and the other members of my family with their love and devotion kept me happy and able to focus during all these years. My son, Dastan, and my daughter, Sheefteh, have been special inspirations for these dreamy years.

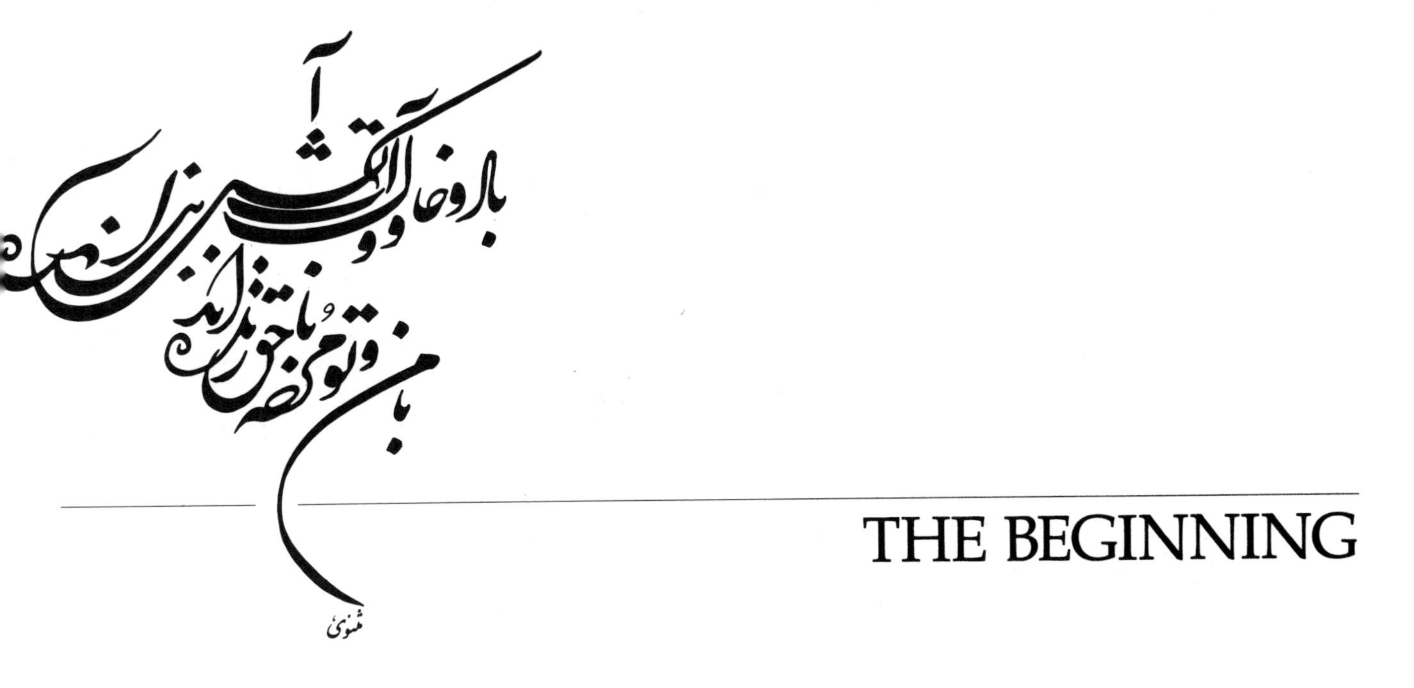

THE BEGINNING

There are twenty-two of us in this bus heading south toward the Navajo reservation. We are a varied group. The students are Americans, Africans, Arabs, Armenians, Canadians, Cubans, Europeans, Indians, Iranians, Asians, Pakistanis, South Americans, and a few mixed-blood Turks, Russians, and others of the East and West; myself, my eight-year-old son, and a ten-year-old sister of a student completes our unique desert caravan. By pure chance we represent almost all continents of the world and all major religions. Except for a student of economics and the two children, they are all my architectural students. Their backgrounds include carpentry, painting, pottery, sculpture, engineering, auto mechanics, music, writing, and other arts. Michael Winter (God bless his soul), one of my American students, and myself are the only ones allowed to drive the bus—and he does most of the driving. We are all bound together by our dreams and the good earth. We are seeking, learning, and dreaming of the simplest and most beautiful form of human shelter, made by human hands from the mother earth.

I have just returned from seven years of research and work in Iranian desert villages, and this is my first teaching work. The manuscript of my desert odyssey is being read by the publisher while I am on this trip, continuing my earth architecture work halfway around the world.

At night we sleep under the open sky in our sleeping bags. During the day we eat our lunch in the bus right out of the grocery bags. We sing songs, argue, read books, or take naps. The Eastern women braid the Western women's hair. As we climb higher, the two students from Cuba and Pakistan experience snow for the first time.

We explore many new ideas and experiences. We see the colossal natural arches and vaults on the Navajo reservation. We wonder why dome-shaped buildings, called hogans, are male or female. We learn the wisdom of building pueblos with earth. We experience the great respect for medicine men in traditional society that is resurfacing in our technological society.

And we touch and test the earth everywhere—the great abundance of the rich clay-earth, a forgotten wealth that could again create the most beautiful homes for the Native Americans. We also see the destruction of the land and culture by government policies and get-rich-quick schemes: the slums built in the name of modern architecture; the social agonies created in the name of civilization, abundance, and freedom; the sufferings in most Third World countries. But the rich earth, the wilderness, and the beautiful cultures keep us going with good feelings and great hopes.

Tsosie, my Navajo-Seminole student, is taking us to his family on the reservation. We arrive late at night, but his family is waiting for us. His mother cooks for everybody, and the twelve women crowd into the small kitchen trying to help. After a few minutes his relatives and friends join us, and the room is packed with many enthusiastic souls.

Late that night, after dinner, I ask Tsosie's father if he could oblige us by showing his paintings. He accepts happily. We all sit on the floor, over thirty people in one room, while my boy and the little girl fall asleep on our laps. The proud and husky father, a striking figure with his long grey hair and turquoise jewelry, stands up and shows some of his paintings. We are awed by the beautiful colors and intriguing interpretations of his culture through the blazing suns, earth, warriors, dancers, coyotes, and horses—the turquoise horses.

We are tired, but we don't want the evening to end. Three of the Native Americans fetch their drums, and the evening comes alive again. The children wake up and the sound of drums and singing echoes into the night.

We stay in the local high school for a few days and later go to the community college and participate in the Native American Folk Festivals. I am invited to give a slide lecture and hands-on clay workshop with my Native American student. Men, women, children, and my own students building many clay models—domes, hogans, interpretations of Native American pottery and glazing techniques and the native Iranian dome structure. Kisses and hugs and tearful eyes radiate joy. The sound of drums, folk dance, the earth and fire fuse our many colors and cultures into one. A local priest comes forward after my lecture, takes his holy cross from his neck, and puts it around my neck. He is touched by my talk and I am touched by his gesture, and as we hug I know I am doing right. That night he takes us to another priest, Father Tom, a wonderful man living in an old mission. We too stay up late and pour our hearts out to each other. Our group—a Franciscan priest, a Muslim architect, and the students representing Islam, Christianity, Judaism, Zoroastrianism, Hinduism, Buddhism, and Native American religions—sits in an old mission at the heart of the reservation, carrying on conversations that radiate rays of joy and hope. I give the priest my manuscript of *Racing Alone*. He stays up late and reads it, and later sends me a wonderful poem he had written that night about the importance of racing alone:

> There is a time for a few
> when straining with and against
> (grade school buddies
> for a mark or prize or popularity
> or affirmation,
> partners in adulthood
> for wealth or position or time
> or recognition)
> is no longer of great import.
> From hurt leading to introspection,
> others' pain begging compassion,
> anxiety calling forth re-evaluation,
> For whatever reason or feeling,
> Stepping aside from the track
> or just slowing down.
> The straining against becomes being for
> And the price of doing with less
> is paid for the freedom of
> racing alone.

On our travels we share ourselves, our cultures, our ideas, and experience true learning. I even dream of starting a caravan school, moving the world over. Many of us come back from this trip wiser and more determined to walk the path that leads to people, the earth, the elements. The ideas that follow are the result of ten years of searching, writing, lecturing, building, and experiencing that path.

Before we begin learning how to build with the elements, I first must try to answer a few questions put to me by my heart, my students, and above all the people I have met and learned from. Thus I must start with my simple *dâstân*, my tale.

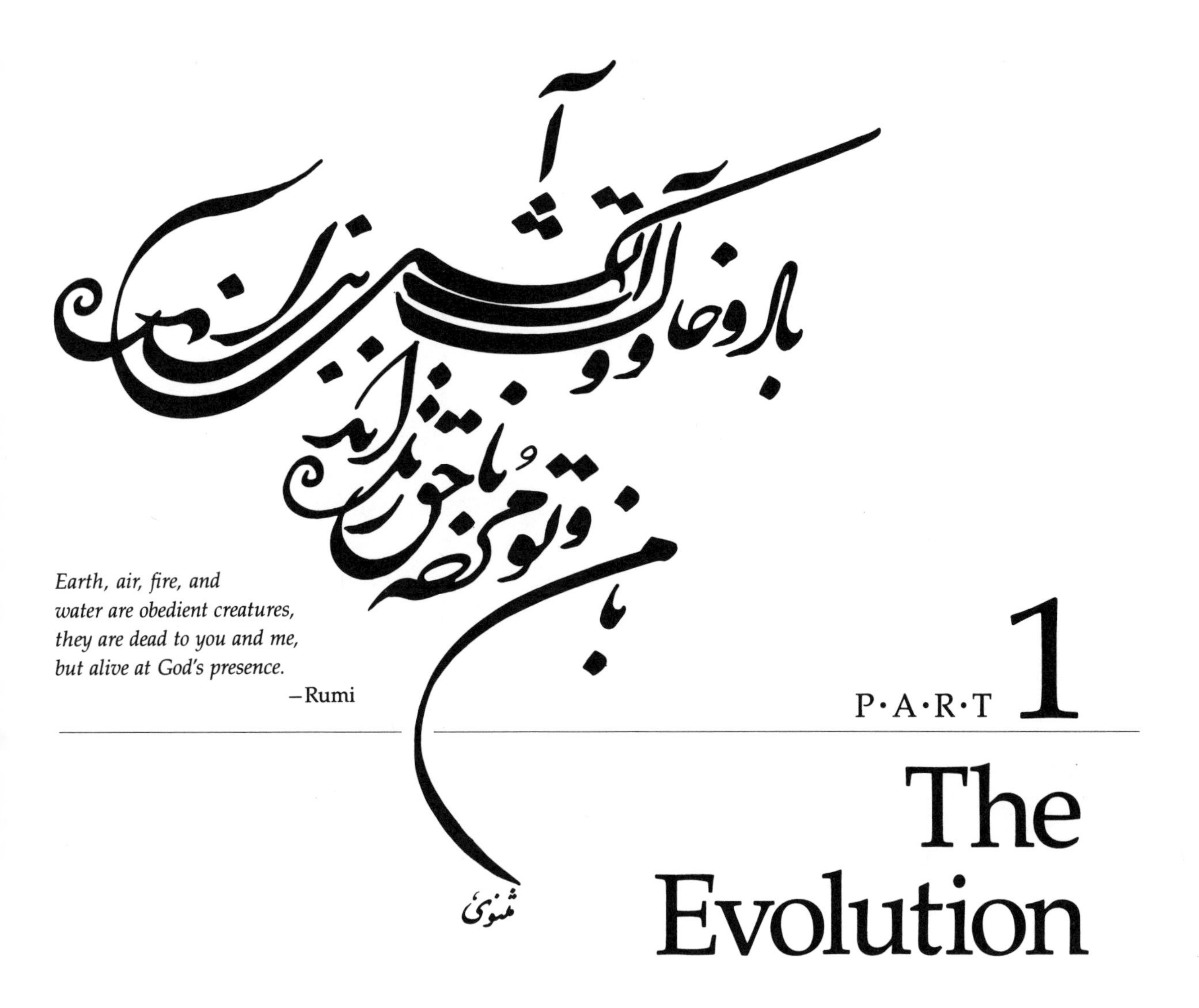

Earth, air, fire, and water are obedient creatures, they are dead to you and me, but alive at God's presence.

—Rumi

P·A·R·T 1

The Evolution

C·H·A·P·T·E·R 1

A TALE OF TWO COMMUNITIES: BOSHROUYEH AND TAOS

Here is a community called Boshrouyeh in the desert of Iran. Here is another community, a pueblo called Taos in the New Mexico desert. The two societies are on opposite sides of the world. They know nothing of each other's existence. Two entirely different cultures, religions, races, and color of skins—and yet, spiritually and physically, their architecture is similar.

The living pattern of each is generated by the people's spirit and the environment. Both look inward—Boshrouyeh around the courtyard, and Taos around the *nansipu* and plaza. Their orientation towards the sun and the wind is based on the same intuitive logic of living in harmony with nature. Earth, the main building material, is the most available and suitable insulation against the harsh desert environment. Neither community is built to last forever, even though both follow the philosophy of permanence—that is, to live in harmony with the permanent natural forces. The earth buildings in Taos symbolize the "rising-from-the-ground" spirit; those in Boshrouyeh represent "dust-to-dust," the return-to-the-earth spirit.

Both peoples have built with the highest intuitive design sense, creating the most compact structures using the maximum common walls and minimum exposure to the outside; in the same way, nature builds its structures using the minimum material for the maximum space.

Both communities use curved forms, Boshrouyeh for its roofs and Taos for its bread ovens. Taos had the timber of the trees and thus chose the flat roof; Boshrouyeh had only the earth and thus developed the curved form to its furthest limits. Taos used the trees to build ladders, and climbed to the higher flat roofs for safer and better living; and Boshrouyeh built taller and thicker walls for safer and cooler desert life.

The only messengers traveling from one community to the other were

1.1 Boshrouyeh, Iran.

the sun, the moon, and the wind; and the only riches common to each society were the earth, the other elements, and human intuition. Both communities have served their people for centuries, and even today—technology notwithstanding—they remain the most suitable design solution for their environments.

Why, then, are we trying to break the **adobe*** walls and roofs of Boshrouyeh to replace them with steel, glass, and plastic? Why do we want to conquer Taos with cheap mobile homes and tract houses? Are we, so-called progressive humans now—a thousand years later—growing senile? Why is it so difficult for us to keep building simple, yet profoundly appropriate architecture?

Our messengers, the sun, the moon, and the wind, are still moving the same route; and the earth and elements are still as abundant as before. They haven't changed; we have. We must have taken a wrong turn in history and gone astray. The rationalizations we offer—today's more

**See the Glossary at the end of this book for definitions of boldface terms.*

1.2 Taos Pueblo, New Mexico.

complex socioeconomic problems, larger population, and demands for more comfort and "better quality of life"– are no justification for our misplaced values. Even if we disregard the lessons of history and view the past as old and decayed, at least we should believe our own "modern" logic, which shows that such communities as Boshrouyeh and Taos have chosen the most appropriate design and material for their environment, and could easily be built today. Instead of desperately producing and pushing new materials and designs, shouldn't we be working on the *right* concepts of economics and the "better quality of life"?

The very least we can do is to raise our level of knowledge and consciousness to safeguard such communities until the "modern" world catches up with them. And if we are to use our technology to teach, then we can also learn from Boshrouyeh and Taos instead of from our factories and our salesmen.

I would like to share with you my knowledge of Boshrouyeh, in particular, to point out ways we can continue and improve on the best traditions of architecture.

A TYPICAL DESERT COMMUNITY

The small town called Boshrouyeh sits near the edge of the Dasht-e-Kavir desert in Iran. Its typical, indigenous architecture, which has evolved over centuries, is created from the earth. Its forms and solutions live in harmony with nature instead of seeking to dominate it, as we do today.

Through the centuries Boshrouyeh has had to cope with a harsh environment—hot winds, storms, scorching sun, snow and freezing ice, hot days and chilly nights. Instead of trying to fight these severe condi-

1.3 Private residence with wind catchers, Boshrouyeh, Iran.

tions with air conditioning, steel, or concrete, the people of the town have used the available materials to create a pleasant atmosphere. **Wind catchers** cool the buildings with the wind; **skylights** cut out the glare of the sun and bring in the light.

The people of Boshrouyeh have learned exactly how to build wind catchers and how to orient them so that their backs are to the harsh and hot wind of the desert but they bring in the breeze. These wind catchers cool the house for the entire life of the building. The wind may be used just as it is; or it may be brought down through the basement or passed over a small pool and fountain or even wet bushes, to create an evaporative cooling system.

Unfortunately, many of these ancient wind catchers have been replaced by big and inefficient water coolers. Unlike them, however, the wind catchers work on many levels—as exhaust systems, as wind suppliers, as orientation towers, and as beautifully sculpted forms.

The courtyard is another important architectural element. The shade side is used in the summer, and the sunny side in the winter. These wind catchers and courtyards embody the philosophy of living with what is offered by nature.

THE PHILOSOPHY OF PERMANENCE

The whole town of Boshrouyeh, like hundreds of other towns and villages, is built with walls and labyrinthine alleys that break the wind and cast shadows. Let's explore one element—the wall. What is a wall? How is a wall perceived by Eastern cultures? Walls of the type we find in Boshrouyeh could be built by two people using a single shovel and some water. For kilometers they could continue to build the ground up into a wall 1 to 8 meters (3 to 25 feet*) tall, using nothing but the earth. We may ask, "What is the advantage of building walls? What is the use of so much wall?"

If we look at the West and its literature, we can see that it has always tried to tear down the walls. "Don't fence me in," as in the American folk songs or the beautiful poems of Robert Frost: "Something there is that doesn't love a wall. . . ."

But in the East, walls have an entirely different meaning. The Persian word *hamsayeh* is like *hamdel*—with the same prefix—which means two people who are very close to each other, or are united in heart, and *sayeh* meaning shade. Neighbors are called *hamsayeh*, people who share the shade of the same wall.

So a neighborhood starts with a wall, united in shade. Each side can take advantage of the shade—one in the morning and the other in the afternoon. In the East a wall is built to break the wind, to create privacy, and to have the shade.

**Measurements given in parentheses are approximations of metric measurements. A metric table is included in the Appendix.*

1.4 An abandoned, two-century-old ice reserve system, rehabilitated with the help of the last generation of ice makers by the Geltaftan Group in Rey, near the city of Tehran. The mud-pile wall (*chineh*) is constructed from the soil dug out of the adjacent ditch. The wall casts a shadow on the north side over the ditch and in the winter, when the ditch becomes filled with rain and snow water, ice layers are collected in the shade of the wall. Ice blocks, cut and floated over the water, pass through an access opening in the wall and are dumped into a covered underground pit on the south side of the wall.

I have built a wall in the desert.
A tall wall.
A wall five knees high.
It is casting a cool shadow on the scorching land.
The walls grow much faster than trees here and give better shade for less water.
With a thousand buckets of water I can build a tall wall to give me shade for a thousand years. Each bucket of water in exchange for a year of shade.
And the earth of the wall is the token of land.

These same walls give the city ice in the summer by making ice and protecting it in the shade in the winter. Here's how it's done. In the winter, they would fill the shallow ditch on the shady side of the wall with water; and every night or cold day, as the tall wall would always create shade and cool breeze, there would form a small layer of ice. Every few

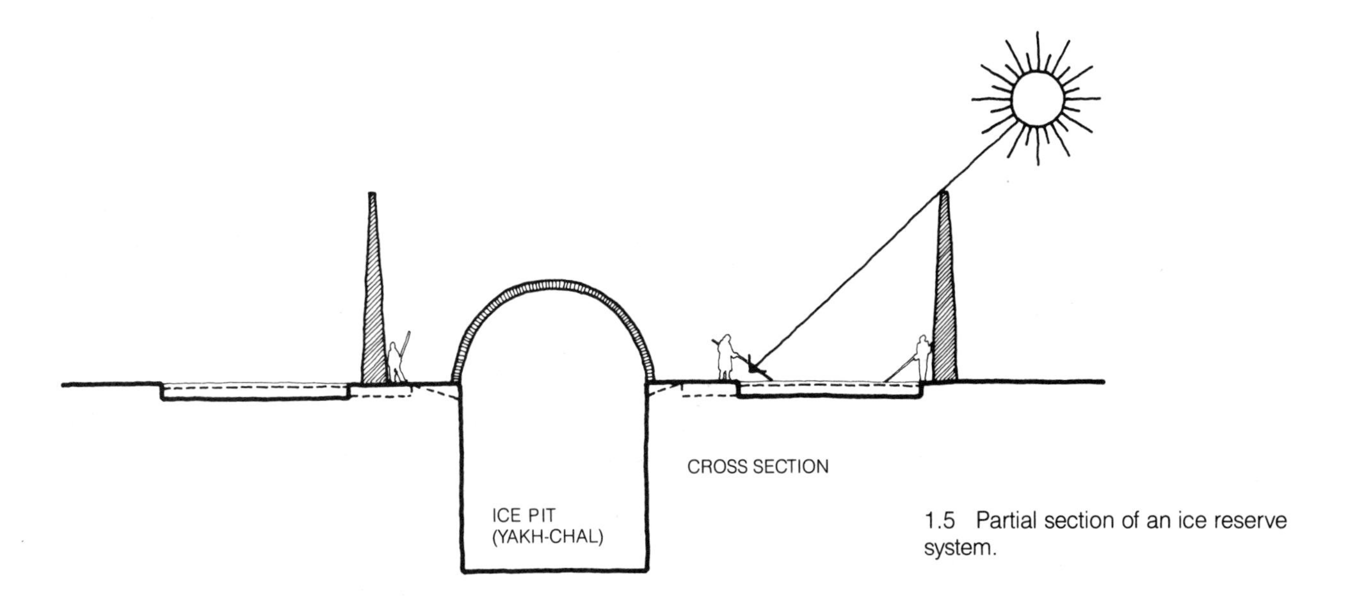

1.5 Partial section of an ice reserve system.

days they would collect it into blocks. Two people, each one with a 3.5-meter (12-foot) hook, would stand one on each side of the ditch to break the ice and float the blocks over the water into the pit. The underground pit (which is around 8 meters [25 feet] deep and the same distance wide, and sometimes over 30 meters [100 feet] long) would be filled up with tens of thousands of tons of free ice or snow. It would then be closed and left for the summer, or even for the next summer.

This is a wonderful example of the use of the fundamental and perpetual elements of nature, designing with permanence, and using an **appropriate technology.*** This is real recycling of energy.

In modern societies, one of our most expensive uses of energy is for cooling or making ice. To make heat we can just start a fire; but for ice we need to make electricity and create the equipment, and then use the electricity and the equipment to get ice or cool air. But the people of Boshrouyeh were using an energy source that is free and abundant: Instead of using the energy of the sun, they used the energy that exists in the shade of the sun.

Today we can utilize this ancient ice technology to make cold storage houses that don't use electricity or freezers, or for a building cooling system. A similar system that utilizes the potential energy of ice is currently being developed at Princeton University. I think they could take a shortcut by visiting and talking to the people of Boshrouyeh. Just think of it—

**Qanats*, *the main water irrigation system in Iran, have been in use for several thousand years. This very simple system of underground canals brings in water for miles from higher ground, where water is available, and lets it flow to the surface. A distance as long as from here to the moon of these canals is still working in the country today. Each one is built for permanence, and many have flowed for over a thousand years. Our modern deep-well pumps, on the other hand, give us trouble every few months and can dry up the reserves or lower the water level. Qanats also generate communities that grow according to the water afforded by nature, rather than according to forced and variable water pump supplies.*

cities like Chicago, Tehran, Paris, or New York could take all the snow and ice that creates problems in the streets during the winter and preserve it to cool their buildings in the summer!

Thus we can see what walls can do. So many towns have been built with these walls. Some of the walls are 300 meters (1000 feet) long; six of them have two ice pits. That would give maybe twenty thousand people enough ice for the summer. In Part 3 we discuss the techniques of how to build **mud-pile** walls and how to reserve ice.

Now this is what happened with the coming of industrialization: Modern architecture started in Tehran by tearing down the walls, as a symbol of progress and Westernization. This was the beginning of what was "modern." And once we tear down the walls, we tear down the whole culture that goes with them. The privacy, the shade, the windbreak. Then we come to rely on ice-making machines rather than the eternal walls—the temporary and expensive rather than the free and permanent.

The high-rises are built right on the sites of the old walls. These high-rises fit none of the criteria necessary to justify them. They are more expensive, harder on city services, and create land speculation. But I don't want to talk about these problems. I want to talk about the good things, like the earth.

C·H·A·P·T·E·R 2
EARTH ARCHITECTURE

We can build a home using nothing but earth, either by making adobe or just piling up clay. Adobe is a sun-dried block made of earth. With this adobe and the knowledge of how to build **arches**, **domes**, and **vaults**, we can construct a building. We don't need steel, we don't need concrete, we don't need any manufactured materials or even timber to cover a roof. There are many examples around the world to show us the strength of these local materials and techniques. One of the main reasons for the invention of arched forms was the lack of timber. The necessity created the form.

Anybody can learn to build arches, vaults, and domes. Hassan Fathy, the great Egyptian architect and the pioneer of modern adobe architecture, learned from local masons and started building and teaching how to build these forms several decades ago. We can build with small pieces of material such as adobe, instead of large members such as timber or steel. Even with small pieces of rock we can create structures that will last hundreds of years.

A dome starts from the four corners of a square room. We cover up the room very simply, without using a form, by leaning small adobe bricks on top of each other. The system is appropriate for many parts of the world. In places where people are struggling to get lumber, steel, or similar materials to cover their roofs, techniques such as these are ideal.

It is also simple to learn to build a vault. We start by building a leaning arch from one end and continue building rows of leaning arches. The simple spans of around 3 to 4 meters (10 to 14 feet) could be learned—by ordinary people, and, of course, larger spans can be built by master masons.

Some engineers may argue that we really cannot use adobe and clay beyond a very short span, or beyond the elastic limit of these materials.

1.6 Adobe blocks are used as the form-work for an adobe arch and removed afterwards.

Yet very large adobe spans are still standing after five hundred years. Why is there such a dichotomy? Because technology give us only piecemeal knowledge of individual materials, rather than the relationship of the elements in nature.

With that same earth, that same clay, we can build many forms. We can build many forms for many functions, without using any equipment, tools, or **form work**. Look at this window, the lintel arch. See how it is formed? Adobe has become its own form work.

The issue of passive heating and cooling is very popular today. Here too, the arched roof deals with the sun very efficiently. A flat roof is exposed to the sun all day long. But a curved roof always has a shade zone and a sun zone, which creates two different temperatures and thus a draft.

I have seen some interesting examples of what happens to these adobe or mud-pile buildings after a disaster. Many of the houses built by the government were of concrete and had flat steel roofs. But the people tore some of these roofs down and built arched adobe roofs themselves. They said it was cooler in the summer and warmer in the winter. Very simple logic—and yet it was not understood by the technocrats.

Many of these beautiful old earth structures are sliced right through to create parking lots and boulevards. The officials involved have no idea as to how these traditional forms were constructed or used. Just for the sake of progress, the old must be destroyed to open room for the new and so-called modern!

EARTHQUAKES AND EARTH ARCHITECTURE

Towns and villages built with earth are not without problems. Sometimes an earthquake destroys an entire town. The town of Tabas in Iran suffered an earthquake measuring 7.7 on the Richter scale. Practically the only things left standing were rug frames and gates. But then, the modern concrete and steel structures were also destroyed.

Now I am not saying that we cannot build steel or concrete structures that can withstand earthquakes. I have been involved in designing and building many standard earthquake-proof buildings in California. What I am saying is that, to several hundred thousand towns and villages in the world—to 65,000 villages in Iran alone—we will never be able to take architects, engineers, steel welders, concrete. With the going rate of building, it would take thousands of years to do that in Iran alone.

What happened to the steel frame building in the Tabas earthquake was this: Plenty of steel was used, but the workers in most cases did not know how to weld it. So the whole thing collapsed. That applies to many of the high-rises in the modern cities of the Third World. Buildings are professionally designed and stresses calculated by computers, but they are welded by kids who can barely stick one piece of steel to the other. The material alone has no meaning; it must have the appropriate technology to work.

We cannot buy the technology in the marketplace, as we buy a bag of cement or a piece of steel. Once a technology is imported or exported, its implementation in the new place becomes very important. When we talk about appropriate technology, we also mean the people involved in its implementation.

It is also necessary to differentiate between a "big" earthquake and just any earthquake. Every year hundreds of shocks pass through the adobe buildings of the world. They either ride out the shocks safely or develop cracks, which are promptly repaired with mud. It is only when a "big one" hits that the disaster starts. And a big one is what they had in Tabas. Comparing the 7.7 magnitude of Tabas with the 6.4 quake in the San Fernando Valley, California in 1971, we can see how destructive both are, regardless of construction techniques. But at Tabas, the old brick water reservoirs, several of them over a century old, withstood the earthquake; the modern buildings did not. Many adobe and mud structures, especially those with domes, also resisted well.

The argument about the vulnerability of earth architecture to earthquake is one of the greatest weapons used against it. Even though no one can deny the low seismic resistance of adobe structure, the argument itself is more destructive than the earthquake. And instead of improving on its seismic resistance, almost all efforts are directed to replacing adobe with manufactured materials. There are many files, studies, and reports by architects and engineers who have studied the aftermath of the great earthquakes in the mud villages of Iran. Almost all of them point to the survival of the domed brick or adobe structures, and to how they could be the right solution (especially in the harsh climates) if their structures were strengthened. Official reports also tell how some of the newly constructed concrete structures failed the concrete core test, since the compressive strength of concrete was weaker than that of adobe. The construction workers didn't know how to mix the concrete. Many prefabricated steel and concrete housing projects have been put up in a hurry after disasters, but many people ended up using the prefab boxes for their animals. They then built their own buildings with familiar materials and familiar spaces.

But many such points are completely ignored. Reports are left to gather dust, while the technocrats keep prescribing steel and concrete and pressing people in a struggle to find a piece of this or a bag of that. I made the following observation after the 7.7 magnitude earthquake in Tabas.

> To complete the Tabas episode, I decide to take the twenty-hour bus ride back to Tehran. It will give enough time to brainstorm with myself and to complete my notes on my thoughts. The bus passes by many villages and thousands of adobe and clay buildings with their vault and dome roofs hugging the horizon. What keeps my mind occupied more than anything else is the damage of the earthquake in this country; not the damage done by the tremor alone, but that done by the specialists. Every time a quake hits, it becomes happy days for the technocrats, the housing ministers' advisors, and the cement and steel merchants. They keep attacking the traditional architecture, the "uneducated local masons and builders," and the lack of government control on construction. They create such fears in the hearts of the officials that for a long time the officials don't dare to touch anything without these advisors' sanction. The man in the village goes through hell and fire to put together enough money to buy steel beams and bricks. To construct brick walls with mud mortars, to build a steel beam roof sitting on shaky support, to use huge lintels for a giant window facing the sun, and to mount a gigantic water cooler on top of the flat roof to fight the desert summer. The whole work is nothing but a mockery for a culture that has boasted of the ingenuity of its native architecture for centuries.
>
> The myth of the earthquake-proof structures created by the specialists and the government advisors should be uncovered. Sixty-five thousand villages of Iran will wait for

1.7 In San Fernando Valley, California, an earthquake of 6.4 on the Richter scale damages beyond repair a newly constructed, all-concrete hospital (1971).

1.8 In Tabas, Iran, an earthquake of 7.7 on the Richter scale destroys most steel, concrete, masonry, and adobe buildings (1978).

3,000 years if they are hoping to be rebuilt with these technocrats' recipes:

"All structures should be built with steel and concrete to become earthquake proof. Government should not invest in any type of building that does not follow the seismic code, neither should it let the people do so."

I write down these notes in agreement with them: Yes, all the new buildings in 65,000 villages and small towns of Iran should be built as they say, but only—

If there are enough engineers to design them.

If the engineers would be ready to leave their comforts in a city and go to the villages to supervise.

1.9 In the Tabas earthquake, the steel-frame structure, shown on the previous page, fails within a hundred feet of several adobe domes and vaults such as this, which withstand the shock. Most large-span domed underground water reservoirs built with masonry also survived.

If there are roads to these villages.

If there are enough supplies of imported cement and steel after the building of dams and infrastructures.

If these materials could withstand the scorching heat and freezing cold weather to make life bearable.

If steel and cement are not used as the symbols of wealth by well-to-do families to reinforce the social classes of richer and poorer people.

If they would be accepted by the villages.

And if all these were possible, and we were able to replace all native buildings with the so-called modern and earthquake-proof buildings, at the end we would have nothing but a disaster: the loss of a wealth of culture and tradition.

It is amazing how blind these specialists and government advisors are to seeing that every time a large quake hits, all their "modern" buildings are leveled as well, and all that is left, if there is anything, are domed clay or brick traditional buildings, more of which will last if some improvements are made on their materials and techniques. But the advisors have access to the media and the offices of the decision makers, and they have thick books, in foreign languages, to prove their authority beyond the shadow of any doubt.

If we compare earthquakes, floods, fires, and other disasters around the world, including automobile accidents, we can get a better perspective on how "right" it is to abandon earth architecture because of its shortcomings. Wood construction in America and other parts of the world should be totally abandoned because of its vulnerability to fire, and the immeasurable loss of life and property it causes. Automobiles, which probably bring more deaths and injuries than earthquakes, fires, and all other natural disasters put together, should also be abandoned. Out of a yearly average of ninety thousand deaths by accidents in the United States, around fifty thousand people are killed in auto accidents alone.

There is no great effort to abandon wood construction or the automobile in the West, but the Third World is willing to abandon earth architecture. "Modern" thought says it is more logical to find fireproofing protections for wood construction and safer design for cars and roads. The main factor is the "profit motive." Earth architecture could not be as easily sold to people as wood, steel, or cars.

But as the Western world discovers the great advantages of **earth architecture**, and invests in it and improves on it (some of the new changes in local adobe building codes point in this direction) it will not be long before the Third World must start buying its own traditional construction techniques from the West–either improved, or just presented in more glamorous ways. The Third World now has an unprecedented opportunity to invest in research and thus improve its own appropriate earth architecture, rather than run wild to imitate the already discarded industrialized ways. The authorities in the developing world must demand that their architects, engineers, and researchers come up with earth architecture that is safe, affordable, hygienic, and beautiful.

WATER AND EARTH ARCHITECTURE

Water may be even more destructive to earth architecture than earthquakes. Plain wet weather is the major enemy of buildings built with earth. A big earthquake is like a raging war, while water is like a chronic disease. Earthquakes get the big headlines, just like plane crashes; but except when it becomes a flood, water is a quiet killer.

The inherent strength of an adobe arch, dome, or vault becomes zero when the material is wet. These strong structures become liabilities when they start melting in the water. If an earth building is not well protected, it will fall apart with rain or snow.

Straw and dung, rice fibers, animal blood, egg yolks, oil, stones, ceramics, lime, cement, asphalt emulsion, and chemicals are among the materials used to protect earth buildings in different parts of the world. Even today, when new materials have flooded the market, we can build and protect earth buildings against rain and snow using traditional materials and traditional ways. However, new ways to improve the qualities of the materials and techniques could set new traditions if the basic philosophy of earth architecture is followed.

Water, like the other universal elements earth, air, and fire, is both con-

1.10 Snow and rain have destroyed most of the houses built with adobe in this village in Iran.

structive and destructive. When in equilibrium with the other elements, water can create balance. The water that destroys the strong structures of earth could also help create landscapes integrated into the building, or help to cool interior spaces.

The concept of four elements is simple, yet crucial. For example, to a ceramic bowl, which includes three elements (earth, air, and fire) water is a welcome addition. Each element enhances the other, none destroys the other.

In a word, what we lack in our earth architecture may be the fourth element, fire. Fire can bring about an equilibrium with the earth, water, and air. And that thought led me to search for an answer.

C·H·A·P·T·E·R 3

GELTAFTAN (FIRED STRUCTURE) EARTH ARCHITECTURE

> Midway in my life I stopped racing with others. I picked up my dreams and started a gentle walk.
>
> My dreams were of a simple house, built with human hands out of the simple materials of this world: the elements—Earth, Water, Air and Fire.
>
> To build a house out of earth, then fire and bake it in place, fuse it like a giant hollow rock.
>
> The house becoming a kiln, or the kiln becoming a house.
>
> Then to glaze this house with fire to the beauty of a ceramic glazed vessel.
>
> I touched my dreams in reality by racing and competing with no one but myself.

It started first as an inspiration, and then became a dream: to create human shelter out of the four universal elements.

Right in the middle of the building boom in Iran, I folded my successful architectural practices in Tehran and California, bought a motorcycle, and drove out to the desert. I was ready to give up many times during the five years of searching, but too many forces had power over me: God, the search for a meaning in life, deep enjoyment of being involved with common people, hunger for knowledge, my strong educational background in the East and the West, determination, and probably fate. It took five years of struggle mixed with joy—the national revolution mixed with my own inner peace, but always the intensity of the quest—before I touched the first results as buildings, and seven years before I saw my odyssey in print.

When I sat in my workshop at the edge of the desert—three mud walls and a tent cover—I was burning with my dreams. I also had a great love and respect for the native architecture – not the grand monuments, but the people's architecture, particularly those built with earth alone. I started making clay models, copying the existing forms and experimenting with new shapes. With a boy as my assistant, a village family as my neighbors, and lots of time, life was molding me as I was molding the clay.

We built clay models for imaginary, but ideal situations. We built a room model on a sand foundation, fused it with **fire**, and let the "earthquake" shake it as much as it could. We separated our building from the ground rather than anchoring into it. We made small models towards the assembly-line production of building single-vault forms that could be carried to the site and arranged in various combinations. We explored building a house, a school, or any other complex with monolithic vaults. We created a model fired wind tower, which could also collect heat through a solar collector sculpted over it. The clay forms with sawed flower-pot windows and sculpted interior shelves and spaces and many other experiments were encouraged by the firing process. We learned that once a sculpture, a room, or a building is fired, a new phenomenon is created!

After a year of work I discovered master craftsmen working with clay who were many, many years ahead of what I was trying to do. There was Nasser Aga, the man who was making huge bread ovens, that looked like small rooms. They had 4-centimeter- (1½-inch-) thick walls and, when dried could be carried to another town on a truck bed for several hours without cracking. There was Ali Aga, the **kiln** operator and ceramist, who could tell what temperature a fire was just by looking at it. I began to learn their techniques and expand our earth architecture experiments.

1.11 Large-size bread ovens made of clay-straw or clay-dung mixtures. Tehran, Iran.

The major problem was how to build a room and put it in a kiln, or build a house and then construct a kiln around it. I started searching for firing systems and kilns around the world, but I found my answer right where I was, in the fork of a desert village road. There is a Persian poem that says:

> *âab dar koozeh o mâ teshneh labân migardim*
> *yâr dar khaneh o mâ gerde jahân migardim*
> (Water is in the jar but we go around thirsty, the sweetheart is in the house but we search for one around the world.)

In the West we would say, "Diamonds are in your own back yard."

The system I came across was this: Make sun-dried adobe blocks, pile them up in a circular form, like a tower, 100,000 or more blocks, and start a fire in the tunnel under the tower. Simple. No kiln. One-centimeter- (3/8-inch-) thick mud-straw plaster covering the outside perimeter works as the kiln. The fire penetrates all the way to the outside and bakes all the blocks. I realized we could bake our houses just like this, fired from inside out! The building becomes its own kiln. Adobe block walls and roofs have no vertical **mortar**, and thus fire can penetrate to the outside plaster covering. Pictures and explanations of this no-kiln system are given in Part 4.

A TYPICAL VILLAGE EXPERIMENT

Our first chance to implement this technique came to us not as a new construction, but as a rehabilitation work of a village's old housing. This village, Ghaleh Mofid, was typical of thousands across Iran and neighboring countries—houses built for farmers and their animals from the earth alone, dug out from the site. The sixty families who used to live in this village either left or were harmed by cave-ins of the mud roofs. Twelve were two-room houses still standing, and twenty-five were partially or totally ruined.

The twelve families living in the surviving houses became our clients. These clients had no money to give us; they could only afford part of their time and the abundant earth around the village. They were also ready with their prayers and their bread and yogurt, to help us do our work. And except for one or two who stayed suspicious as a matter of habit, we had the fortune of winning everyone else's trust and cooperation. Our small budget came partially from the new provincial government and a private organization, and partially from our own pockets.

Ghaleh Mofid, like many other small villages, is almost unknown to outsiders. The typical house was built with a single-vault roof and a low partition in the middle to divide the one room into two; there were no living room or bedroom divisions. In this part of the world, the rooms are not divided into living rooms, bedrooms, and dining rooms. All rooms are for living, sleeping, or dining; they are built smaller or larger. There was no electricity or plumbing. The village houses were infested with vermin and mice, and even the drinking water was contaminated.

We approached our work with a philosophy we religiously tried to adhere to: We would make no compromises in our quest for the simplest and most basic solutions—solutions that would employ materials and techniques that could be used anywhere in the country. We were also hoping to reach out to similar bare-minimum-existence villages in other parts of the world.

We found a local mason who repaired the damage done by rain and snow, and helped get the first house ready to be fired. He cut a few vent holes and a fire-pit under the direction of yet another local man, who had the native knowledge of how to fire brick or ceramic kilns. We took out the door and the window of the house and built a mortarless temporary adobe partition in the openings.

At this point, an amazing idea presented itself to me. The house itself was becoming a kiln; and if we could fill this kiln with sun-dried adobe blocks, tiles, water jars, and flower pots, we could fire and bake them along with the house. Then we could use the products for courtyards and walls, or sell them in the marketplace. Thus the house could become a producer as well as a consumer of building materials. And if the process is done right, then every time that we fire a building we can bake brick, tile, pottery, and ceramic products with it; sell the products; and pay for the house. The dream of making no-cost housing instead of low-cost housing with earth architecture could become a reality. We discovered that every time a conventional pottery kiln is fired, around two-thirds of the fuel is wasted in baking and rebaking the structure itself. In a simple calculation, we determined that with the wasted fuel of a modern brick kiln we could build and fire several thousand two-room houses—because in this process every time we fire the kiln, we keep the kiln as our house.

First we fired one of the houses in the row of old dwellings. This house was damaged and was only used as storage. Our team fired and finished only two of the village houses, as prototypes, and the villagers did the other ten structures on their own. They used the same equipment for the firing, but paid for the fuel and replastering from the budget given them by the governor's office. Each house used the equivalent of $52* in kerosene to fire its two-room, approximately 30-square-meter (300-square-foot) area.

Our team, the **Geltaftan** Group, consisted of a kiln operator (Ali Aga), two architectural students (Mahmood and Ezzat), an engineer (Sedehi), and myself. The mason (Ostad Ghodrat), my young sister (Nahid, a social worker), and others also participated. The villagers and other volunteer workers were all behind the success of the simple, yet important work we were involved in: the firing of an old adobe house.

**All dollar figures are in U.S. dollars.*

1.12 The door and the window are removed and the openings closed temporarily by adobe blocks without mortar. The two-room house, which has become a kiln, contains clay products to be fired along with the structure.

1.13 One of the two homemade gravity-flow kerosene burners used for all the firings in the village of Ghaleh Mofid.

THE FIRST FIRING

Our firing technique and equipment consisted of the basic kerosene **gravity-flow** system used and tested by native potters and kiln operators. The steel burner was made of two pieces of sawed pipe welded together, with an air sandwich between them. It was built by a local blacksmith at Ali Aga's direction. Ali Aga was originally from Hamedan, a city of Iran known for its ceramics. Like other potters he had been making and using this homemade torch for many years.

With two barrels of kerosene oil on top of an adjacent roof and connected to the burner with pipes near the ground, we had enough head pressure to start a great fire. And as long as we had enough kerosene in the barrels, the fire would continue nonstop. No pump, no electricity, and no complicated control system—just a valve to control the flow of the oil and thus the amount of fire produced.

We fired the first house from underneath for twenty-four hours, just as if we were baking a pot of rice. The first experience was mixed with great excitement, fear, and miracles. One miracle happened just an hour after firing. We started the torch slowly. Around an hour later we heard the screaming and laughing villagers, with sticks and shovels in their hands, chasing and killing hundreds of roaches and dizzy mice that were running out of the firing house—"Revenge, revenge!" With the heat and smoke, all the rodents were either burned inside or were running for their lives with people in hot pursuit, fired by the remembrance of all the wheat, blankets, and clothes they had lost to the vermin. In those moments the excited villagers praised us—not for our architecture, but for what the fire was doing to their house. The disease- and rodent-infested house was becoming clean, and what the poisonous insecticide sprayers hadn't been able to do, the fire was doing: creating a completely hygienic environment, without the use of polluting chemicals. The element of fire now found a new dimension for us: a hygienic medium. We suggested that they fire their storage buildings and animal shelters once a year, to get rid of disease and vermin. One farmer even thought of moving his own furniture out once a year to refire the house, just for an hour, to keep it hygienic.

As we started the slow fire, the steam began coming out of the house from the flues as well as the roof. For ten to fourteen hours, steam rose from the old adobe structure. Tests showed 17 percent water in the old adobe and mortar material. The inherent water of the **earth-clay** mixture will only come out after the material is heated to several hundred degrees. And since the atmospheric heat on earth never reaches that high, there is always enough water in old earth buildings to be fired to brick, as long as the material is suitable for firing.

After the firing and two days of cooling, we opened up the house. It seemed as though a miracle had happened. The old and crumbling mud structure had changed gradually from a solid brick shell inside to a less burned adobe outside. The owner of the house began digging with a pick from within, to test it, while his wife and children sat on the roof drilling with long donkey nails to see if it had really become solid. Since a

1.14 Steam rises from the old adobe structure for many hours as the fire continues to burn inside. The firing process also makes the vermin-infested house hygienic.

nearby house had collapsed under the rain a week before, their joyous cries of "It is solid, it is become brick," became more meaningful. Even though we had predicted this result, it still seemed unbelievable when we were touching it.

Our greatest reward came when a farmer's wife arrived with her incense tray and praying lips, saying, "Now I can sleep in the rainy and snowy nights without fear of the roof collapsing and killing my children and my man. It is brick. It is mud no more." This was the greatest reward for me personally, since my quest and her need had met: A safe house for her, and a dream reached for me. And we all were happy to be rewarded with the knowledge that what we were doing was right.

The most important part of this work was the people's acceptance of it. At the beginning we had a hard time convincing everybody in the village to let us do our experiment. One of the greatest factors in rural refusal of government housing has always been the superimposition of new materials and techniques. But what we were doing was in reality neither new nor strange to them. They had already seen the end result before letting us try it on their houses. An accidental discovery, or what I believe to be an act of God, helped me break this barrier.

AN OLD KILN: LESSONS OF HISTORY

One day, several hundred steps from the village, I discovered an old and abandoned structure. This structure was used as a **kaval** kiln. **Kaval** is the Persian name for the short pipe sections used in the traditional underground aqueducts, known as *qanats*. They used to fire clay *kaval* in this kiln until a few decades ago, when concrete and plastic pipes flooded the market and the kiln was abandoned. There were hundreds of these kilns around the country, and almost all of them were dead. This kiln structure looked like one of the village houses—a long barrel-vault room. This room was about 7 meters long, 3.5 meters wide, and 3 meters (22½ × 11½ × 9½ feet) high. The kiln was made out of adobe blocks and clay mortar and was baked along with the clay pipes during the firing processes.

We took all the villagers to the kiln:

The several-hundred-step march is like an exodus from the village. To me it is an exodus from the present to the past.

We reach the kiln. Children climb all over the roof and the walls.

"What was this building used for before?" I ask the old man of the village aloud, while everyone is trying to guess.

"It was a Kaval kiln. In my childhood time I saw it fired," says the old man.

"How long ago was that?" I ask.

"Oh, maybe forty or fifty years ago, maybe even more. I don't know how old I am now." He laughs as he says that.

I let them play around with the walls and touch the rocklike pieces.

"They used to fire it right from underneath on this big hole. They used to burn wood, animal dung, or anything they could burn. Yes, see, right around the firepit the soil is melted to rock," the old man says while he tries to break a piece but can't. A younger man kicks a piece with his boot; he can't break it either. Everyone laughs.

Then I stop them and ask them in a low voice, facing the old man and trying to have them observe silence. "Amoo, why have all your houses collapsed but this roof hasn't collapsed? Yet you all plaster your roofs every year and you say that this roof is just left under the rain and snow for thirty years."

"Not thirty but fifty years," he says.

"Okay, fifty years. Why is it still standing?" I ask.

A middle-aged peasant answers in a loud voice from behind: "Don't you understand? This is fired and baked to a rocklike brick. Even a cannonball can't break it."

Then there are a few seconds' silence. Several have already made the connection. My architectural students and engineer friends make the connection first, but before they start to explain what I am trying to say, someone in the crowd says, "So this is the puzzle?"

And in a few seconds everything falls in place.

The history connects with the present. Moments link, and the chain is completed. There is more silence, and everyone is digging a piece or climbing to the roof. More conversation, more comments, and more photographs, even several group portraits for the memory's sake are taken on the roof.

By the time we walk back, there seems to be no question as to the validity of what we will be doing. And everyone offers his own house for the first firing.

So the idea of firing their houses was easily accepted, since no new material or strange processes were involved. Some of the people could easily have lived in that kiln, which had the same familiar form and space as their houses. They were also very familiar with the idea of building with earth-clay (*khak-e-ross*, in Persian), as they dealt with the earth all the time. And even though they hadn't done any firing themselves, the elders had seen it done and the younger ones had experienced firing in brick or pottery kilns.

1.15 A *kaval* kiln, near Ghaleh Mofid village, was used in the past to fire clay aqueduct pipes (*kaval*). The kiln is a room similar in size and shape to village houses. The kiln has been abandoned for decades. There are hundreds of such kilns all over Iran.

Kilns are abundant all over the world, and many of them are built with earth-clay. Kilns of China, Japan, Korea, Nigeria, and the West, including the Native American kilns and bread ovens, are based on the same principle: The earth-clay kilns are fired along with the product inside. We can see many of these kilns—like strong houses, built with earth and fired to last a long time. People live across the street from the old kilns in flimsy shelters that fall in the rain, break with a small wind or earth tremor, or even catch on fire, while their kilns keep living for decades or even centuries. These kilns are time-tested lessons of history to be learned from and followed.

As we continued searching for more lessons from old kilns, we were amazed to find that the interior plaster finishes had also lasted for decades, even though they had been exposed to the harshest weather. The simple **mud-straw** plaster, fired in place, had become a durable and attractive finishing material.

Once the houses were fired and the villagers knew that they would not disintegrate in the rain and would have a more lasting life than their old houses, they started plastering and decorating the interiors according to their own tastes. The elimination of roaches and mice was an added bonus to encourage them to take better care of their living environment. Some of them paved their porches with baked bricks that were produced in the firing process, and even began dreaming of the future when they could own the house and work on their courtyards. And even though such great problems as contaminated drinking water, land ownership, health care, and job security were still unsolved, their enthusiasm and hard work to rehabilitate their houses greatly improved their living conditions.

As far as our own work was concerned, our first project was far from perfect; but we were confident that there were no problems that could not be improved by more time and experimentation. Yes, we could fire a structure; we could even **glaze** one.

A SCHOOL BUILT WITH EARTH AND FIRE

While working on the Ghaleh Mofid village rehabilitation program, we were asked to look at the possibilities of building an elementary school in the village of Javadabad. Before meeting to discuss terms with the client, we went to the village. There we found a local mason who, up to fifteen years ago, had built many earth buildings with arches, domes, and vaults. He was now busy building a "modern" house: imported fired brick to construct walls with mud mortar, huge Swedish steel sash windows, flat roofs built with Japanese steel beams and lintels, Italian PVC plastic pipes, Portland cement finishes, and a German-American water cooler.

We asked Ostad Asghar, the mason, if he could still build adobe buildings with vault and dome roofs. He said he could, but didn't any more, because people wanted "modern buildings." We told him about the pos-

1.16 Interior of one of the houses in Ghaleh Mafid after the farmer fired the structure himself and the surfaces were plastered by a local mason.

sibility of the school project for the village, and he was interested to be the subcontractor for the labor and the material—but he wanted to do it the modern way, not the traditional way.

"Adobe and mud are old-fashioned and people don't like to build with it, even though the houses are comfortable to live in. But city people don't use it and the villagers don't want it either," he said. We didn't tell him about the *geltaftan* system, but asked him if there were any old kilns nearby. He pointed to the other side of the village and said that there was one there. By now many villagers had gathered; and as we started walking towards the kiln, with the mason in the lead, they all followed. The episode of the villagers' march in Ghaleh Mofid to the *kaval* kiln was repeated here, and they discovered by themselves the "secret" of building with earth and firing it in place.

Excited by such an idea, Ostad Asghar volunteered to build the school for one-third of the cost he was charging to build conventional buildings. He asked us to do the firing and accept the general contracting responsibilities. We shook hands, returned to the city, and signed a contract with the development department of the provincial government of Tehran.

We prepared the design and plans, submitted them to the department, and were awarded the contract. The design was for adjacent boys' and girls' elementary schools with common administration facilities. But later, as the work was completed, they decided to use the building only for the girls' school and to make the surrounding courtyard walls taller.

We developed the design with the mason's help, especially the dimensions, since he was more experienced than any of us with earth structures and he was also a native. Later, when the construction started, we found that he could not read blueprints. So we drew lines on the ground in small scale and discussed the details with him. He took with him our notebook page, which had the dimensions. He then put that page, with his own markings, in his shirt pocket, and used it as his "blueprint" to build the entire school.

Ostad Asghar, with the help of the village adobe maker and others, including his children, built the school as he had promised. But he couldn't keep the cost as low as he had promised, since there were many delays—the land-deed problem, which made us move to a new location after the foundations were already built, the crises that arose as the result of Iran's ongoing internal revolution and the war with Iraq. He raised his cost 25 percent. He had also underestimated the cost at the beginning, because he hadn't built adobe buildings for many years and was surprised by the "lazy worker," as he put it. We accepted his demand as a fair addition, but the client didn't raise our budget.

With all this, we still built the final project for one-third of the cost of conventional buildings. Why? The construction involved only a single-trade system: no carpenters, no steel or concrete workers, just a mason to do everything. Also, we used only available material and local labor. We started and finished our school, with all the delays, within a year, while a nearby school waited four years just to receive imported materials. Our group of four (two students, an engineer, and myself) worked for free, but with the total acceptance of the project by the client. This meant that after a year of occupation and correction of faults, we ended up with some cash—which we promptly put into the Geltaftan Foundation.

Many students were trained at the site, including those sent from far-off towns and villages. We also gained experience in building and especially firing, since we had to do most of it ourselves. Ali Aga fired the first classroom; but then, for many reasons, he could not come to the site. (It was far away from his work, and his wife had asked us to discourage him from coming to the firing because of the danger of the war and bombing. We had to keep the fire going at night, even though there was a blackout order. The fire coming out of the roof vents would light the area, and the nearby town had already been bombed.) So we were forced to learn to do all the firing ourselves.

The plan of the ten-classroom school was like a T with long arms. The arms were the classrooms and the stem was the administration section, with two courtyards on both sides. A continuous domed but open arcade connected all classrooms and the administration area. The toilet facilities were built out in the yards, as is the tradition. We only used domes to cover the arcade and vaults to cover the classrooms. The principal's room was also domed. The school, about 500 square meters (5,000 square feet), was fired like a kiln, from inside. First one classroom was fired to test the result; then several rooms were fired at the same time. Different sections were fired and baked, and the adjacent spaces were dried out by the heat of the fired rooms. All together, in actual construction time, it took four months to build the structure and around ten days to fire it. The firing was done in the period of many weeks, however, as the oil was rationed because of the war and other crises. With more experience and facilities, the entire building could have been fired in twenty-four to forty-eight hours.

The roof was covered with mud-straw plaster, as was the tradition. Some of the plastering was done at the tail end of the firing, which made it bake, to a certain degree, with the roof. The interior was covered with gypsum plaster, although we left some of the ceiling areas exposed. Today, even though it has already gone through a few severe winters with heavy rain and snow, the building is still in good condition. But while we were more experienced because of the previous work, it still did not match the ideal example of what could be built with earth and fire. As our first fully designed, constructed, and fired building, however, we have all been proud of the result and its promise for the future.

It was amazing to see how a middle-aged adobe maker made over 60,000 adobe full blocks for the entire job with a single wooden form and a bucket of water refilled by his son, who was the main helper for all his

1.17 Javadabad Elementary School under construction.

work. It was even more amazing to watch the mason build the entire school with his bare hands, avoiding even simple tools. There was no form work or centering; he didn't use even a standard **adz** to break adobe—he used the hard heel of his hand. One assistant would sometimes be there, or his ten-year-old son would hand him **dry-packing**. That was all. And it again was a miracle to watch what the fire was doing to this structure, and then touch the result.

We brought the earth-clay from outside the village. The earth from the site was rich and appropriate, but the earth we brought in was even better. It was from the sites they used to make brick in a local kiln that had long been abandoned. The cost for digging and transporting the earth was minimal. We did not mix anything with the earth to make the adobe blocks. The earth-clay alone gave the good result. We used small features in our design, such as skylights over the classrooms, to aim a natural light over the blackboard. Since Persian writing is from right-to-left, and Roman alphabets are written from left-to-right, classrooms generally need the light from both the right side and the left side. So we decided to combine them by providing the light from above.

In this project we took advantage of the available budget and purchased a hand pump; and instead of filling the oil barrels on the roof with rope and bucket, as we had done in the first project, we pumped the oil to the roof with the hand pump. This truly was an example of the use of an appropriate technology, which saved us a great deal of wasteful and dangerous work.

During the firing, as in the previous project, we and the villagers both took advantage of the great heat rising from the flues to cook pots of potatoes, sugar beets, and other meals. The shortage of energy because of the political crisis drove us to many ingenious uses of every drop of fuel.

The experiments for the utility lines ranged from using lead-covered wiring in between the adobe joints after the firing, to using outside piping. One of the simple solutions was to provide wood strips where the conduit chase would be needed. During the firing, the wood strips would burn out and create the chases. We also used the rooms as kilns to fire adobe bricks and sculptures. But the freezing winter limited our adobe-making ability. We made many experiments in firing with raw oil and drops of water, which creates a higher temperature output by quantity. It also makes a beautiful fire. But the simplest method was still the gravity-flow firing system.

Each classroom—and the small quantity of brick products inside—was fired with fuel that cost about $54. It was calculated that the cost of fuel used to produce and transport fired brick from the factory to the village would be at least two times that much. And of course the cost of building with imported fired brick and imported cement mortar would have been even more.

The economic justification of what we were doing was a determining factor in our building techniques, but it was not the main philosophy behind the work. The main philosophy was to use the local material and indigenous techniques, and to provide jobs and self-sufficiency while creating beautiful architecture that respected the traditional forms and spirit. And toward that philosophy we dedicated our hearts and efforts.

Here are some of the points we learned from our mistakes in building and firing:

■ Adobe-making time should be scheduled to fit the seasons.

■ Don't plaster the roof before firing, since it will crack when steam tries to escape. It is better to plaster at the tail end of the firing, if safe conditions exist.

■ All important cracks must be repaired before firing. Cracks also may develop during the firing for several reasons: expansion and contraction, faulty dome and vault construction, weak supporting walls and not enough buttressing, overly fast firing and cooling cycle, or wrong type of materials used for blocks and mortars. Small nonstructural cracks can be repaired and lived with, but the structural ones will call for the reinforcing of the affected areas.

■ All materials must be tested before beginning to build the structure. Fire the adobe samples alone and then with mortar to assure good fusion, soak them in water for a few days, and then test them for strength.

■ Many more construction and firing experiments must be done before conclusions can be written to standardize construction and firing details and specifications. And many tests must be conducted before the system is perfected.

■ Wet the walls and ceiling of the fired space after the cooling cycle, to neutralize the lime reaction to humidity.

C·H·A·P·T·E·R 4

A CERAMIC HOUSE– A DREAM COME TRUE

> Now the time has come to create a new scale in the ceramic world, to walk out from the womb of a pot to the space of a room.

This was the climax of our work. The glazing of a house, a ceramic space! We were all burning with the desire to sculpt at least an interior space, and to fire it with a ceramic glaze finish.

Throughout history, humans have always associated ceramic glazing or clay firing with any unit that could be carried with the hands – a fired brick, a jar, a ceramic bowl. Now we would use all the techniques and materials used throughout history, but in a new scale. It was obvious that we needed to learn from the past and move into the future. The ceramic tiles used in the kitchens and toilets in traditional houses, even the wash-basins and water closets, are all made with clay and taken to the fire. Here, instead of taking the materials to the fire, we were bringing the fire to the material. And thus a new horizon had opened up to us.

Ali Aga, the kiln man with forty-five years of experience, was our greatest asset. He could look at a fire and tell us what temperature a fire was just by the color of the flame and his own intuition. He was sixty-six years old, but was as full of zest and anticipation of this new venture as the youngest of us.

We immediately started making our own ceramic glaze solution, with Ali Aga's formulas. We used lots of milled Coke bottles, and native glaze agents. It took us a long time to figure out how to apply the glaze on the walls, floor, and ceiling. We couldn't just dip the house into a glaze solution the way bowls are dipped. We couldn't brush the building, since it would take a long time and create uneven surfaces. The obvious answer was to spray the glaze. But in a village with no electricity or pump

1.18 Villagers are using an insecticide sprayer to cover the surfaces with a coat of glaze. Their faces are protected with plastic bags placed loosely over their heads for the short spraying time.

and spray gun, and the almost impossible task of bringing an electric generator and equipment, we were pushed to a simple and yet beautiful solution: using the farmers' insecticide sprayers. Every farmer had one. And since the glaze solution is a water-base mixture, we could use their bicycle-pump-operated sprayer cylinders.

We did several room-glazings on different occasions. We even did a single-firing experiment; instead of firing the bare bricks and then refiring for the glaze, it could all be done in a single operation. But most of our experiments were with double firing.

Most glazing experiments done over the old plaster and adobe walls developed small cracks and blisters, caused by gravel-sized lime pieces in the existing materials. The lime had been fired, later absorbed moisture, and then expanded. Ali Aga later suggested that we should have immediately sprayed water over the surfaces after the room was bisque fired and before applying the glaze. This way we would have "killed" the lime before its growth. From then on we always sprayed water over fired surfaces before glazing or plastering.

The first experimental room was fired for twenty-four hours and refired with low-fire glaze for twelve hours. And thus we had the first glazed house in the world!

Then we approached a new horizon: the integration of graphics, sculpture, and decorative arts into the already integrated arts of ceramics and architecture. What we did was very modest and basic; but what we imagined we could do had no limits—and our master kiln operator would sanction our dreams and help make practical the applications of the dreams.

We were thinking and experimenting in a small way with salt glazing—that is, to use kitchen salt as the glazing agent. The ceramist who uses salt glazing knows its advantages and shortcomings. For example, one of its main drawbacks is that the salt glazes the kiln before it glazes the vessels. But this shortcoming becomes an advantage for us, since our structure is the kiln itself to be glazed.

The possibilities of sculpting and glazing the fireplaces, bookshelves, even some furniture and tableware became very promising. We even fired and glazed some pieces of ceramic jewelry, heart-shaped necklaces for the woman farmer the same color as that of her house. One farmer helped sculpt and glaze several coat hangers right into the wall. The use of the graphics and arabesque design gave us the idea of firing the geometric design and the holy Qur'an scripts right into the inside of a mosque dome. We wrote words of wisdom, and the farmers wrote their names to be fired and glazed for eternity.

And this was the beginning of a greater vision. The vision inspired by volcanoes, and the message they have been giving us as they belch out molten earth and make cave spaces and sculpted forms: The use of the element of fire to bring into equilibrium the destruction created by the element of water, as in earth structures. To fire coastal cliffs to stop the mudslides and erosion caused by the water. To create habitats for humans on the other planets, or the moon, from the soil of the planet and the fire of the sun, and to apply the same techniques in outer space. And to approach all these ideas, not only for sake of beautiful dreams, but to show the unlimited possibilities that the pursuit of the Elements could bring into arts and architecture, from a human shelter to the vast universe. Some of these visions are described in Part 7.

1.19 Ali Aga, sixty-six years old, is the first ceramist in the world to participate in the firing and glazing of a house.

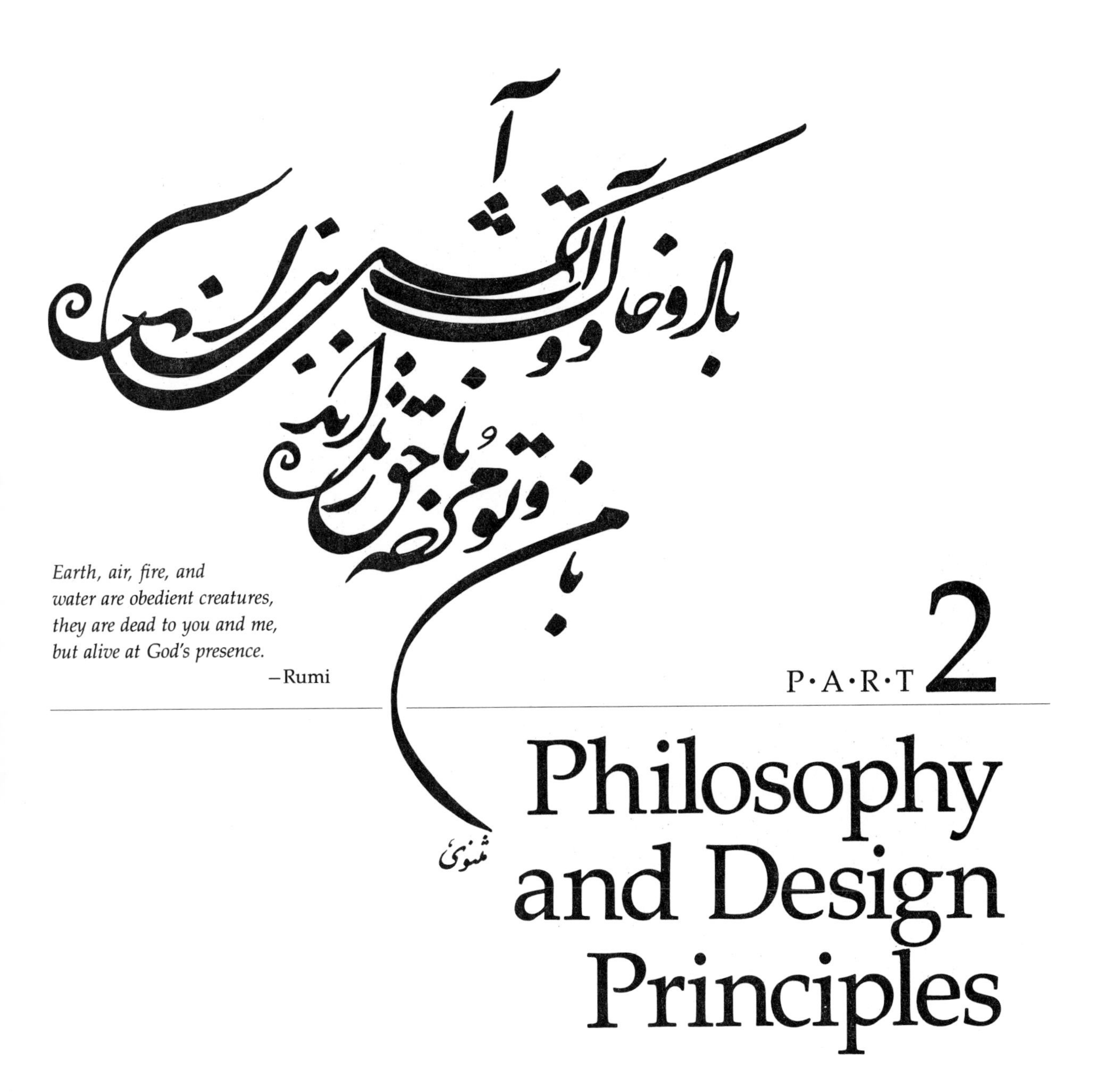

Earth, air, fire, and
water are obedient creatures,
they are dead to you and me,
but alive at God's presence.
—Rumi

P·A·R·T 2

Philosophy and Design Principles

C·H·A·P·T·E·R 5

THE PEOPLE'S ARCHITECTURE: WORKING WITH NATURE

Many volumes are needed to do justice to the indigenous earth architecture of the world. From China to the Middle East and Africa, and from Europe and Australia to North America, a great wealth of knowledge can be documented and used for today's needs. Even though there are unlimited variations in the use of earth for the human shelter, certain basic principles and limitations must be understood. These principles are related to material, climate, design, culture, and more. When we understand these lessons—both physical and spiritual—not only we are not pushed to follow the past, but we continue to improve in the best tradition of earth architecture.

The greatest contribution to earth architecture at this time may well be the documentation and dissemination of the existing know-how of the world's people. We can then draw from the creative ability of all humans, rather than just the few in the building professions, since earth architecture is truly the people's architecture. Let us now explore some of the principles behind earth architecture.

EARTH ARCHITECTURE AND NATURE

Indigenous architecture in different parts of the world has always been created based on the philosophy of living in tune with nature.

> The architecture was created from the earth and, over the centuries, learned to live with the desert: to fend off the sun with its thick walls and long shadows, and to break the hostile winds by labyrinth streets and wind catchers. A city in the spirit of genuine Persian architecture, living in tune with nature with no intention of dominating it. The wind

catchers, backing to the gusts while inhaling the gentle breeze, cool a house for its entire life. The *qanats,* an indigenous Persian irrigation system, let the underground water flow to the surface non-stop for a thousand years. The *yakhchal* keep ice from the winter to the summer to quench the thirsty city with ice water, a true recycling of an abundant potential energy. Amazing: a town built with earth structures at the edge of a desert, having scorching hot days, chilly nights, dusty winds, freezing winters, and boiling summers, and yet creating such architecture that embraces every season, not with fears but with welcoming arms.

The philosophy of living in tune with nature was based on human intuition, which got the work done, and on human logic, which explained it later. The true confusion started to appear when humans began thinking they could control nature. Environmental damage, misplaced social and economic values, and many more problems were a chain reaction caused by the movement of philosophy away from unity with nature. Climatic control is one example of this shift in human thinking and its effects.

For the last several decades architectural and engineering education related to the climate has been limited to choosing heating and cooling systems from one catalogue or another, usually supplied by the equipment manufacturers. Thus architects and engineers have received the best training from the manufacturers of the products and unwittingly become their salesmen. They will go through their normal professional life without questioning the validity of such a route. The basic questions—why use mechanical coolers or heaters? why use air conditioners?—have just begun to bother the public's mind.

It has only been in recent years, since the energy crisis became an important issue in the West, that such questions started surfacing. All of a sudden the sun and the wind started to be felt and noticed again; and architects and engineers did not feel awkward talking about them, even for sleek buildings. All of a sudden millions of buildings, which were designed based on minimum-frontage lots and interesting views, had to be quickly redesigned with makeshift openings to allow the sun to enter and the wind to circulate. All of a sudden thousands of books and catalogues on passive cooling and heating, solar design, and energy efficient buildings were published. Most of these books prove through complex language that the sun is good, the wind is good, the earth is good. These books are good as far as raising knowledge and consciousness is concerned; but they are also bad because, once again, they move toward categorization by the climate-manufacturing industries.

We have seen the most complicated equipment and computer programs employed in the high-cost centers to prove the simplest facts known to every human being—thick adobe walls provide better protection from the sun than narrow concrete blocks. We had the agonizing experience of proving to the Iranian building officials the validity of the use of adobe or brick domes and vaults in the hot and arid climate, even though they see people tear down the flat concrete roofs of the government-built houses and rebuild with their own indigenous vaults or domes. We have seen the struggles of architects to prove to the Egyptian ministry, with costly computer analysis, that the indigenous wind catchers next to the Nile are more appropriate than air conditioning equipment. We have seen how federal housing agencies or banks in the United States will not help Native Americans build with their own indigenous adobe, since they don't fit the codes or comforts—even though the same types of buildings, still standing after centuries, are right there under their noses. Many such events may be plain mockery for the human intellect and common sense, but they are realities that show our loss of respect for our own non-catalogued intuition for living in tune with nature.

EARTH ARCHITECTURE AND CLIMATE

An adobe, **rammed earth**, or mud-pile building with well-arranged windows and vents will not require additional heating or cooling, except in extreme climates and for short periods of the year. The sun's radiation does most of the heating. Additional heating could easily be taken care of by a fireplace, an oil or coal heater, or a simple solar collector system. Since the warm air is retained in the earth walls, roof, and floor for a long time, heat loss is very gradual. Once the interior space is warmed, it stays warm; and with a small heating element, the atmosphere could be kept comfortable.

Since most adobe buildings are built in the hot and arid regions of the world, the cooling system is generally more critical than heating. The use of air conditioning and evaporative cooling units may be an easy way out. But the genius of humans who live in tune with nature has given us simpler, more permanent, economical, and ecologically balanced solutions, which should be followed and improved upon through modern technology.

There are some basic rules related to climate that we must observe anywhere in the world in general and in arid climates in particular.

1. We must use an **appropriate material**, which is generally the earth or other natural materials.
2. We must orient our buildings correctly toward the sun and wind.
3. We must design our buildings with minimum surface exposure to the undesired outside environment.
4. We must make our windows and other openings an appropriate size.
5. We must create as much shade, catch as much desirable sun and wind,

and design with as much water and indirect light as possible; basically, we must design in harmony with nature.

When we have followed all these time-tested, ancient rules, we will have no need for extra help from manufactured equipment. But if we decide that modern technology should be employed, then we must use it to help us achieve the ultimate in basic rules—for example, to catch and store more sun and wind for us in extreme conditions.

Now let us look at these basic rules in more detail.

1. *Appropriate material*. Buildings constructed from earth provide one of the best types of natural insulation against hot and cold weather, if all factors are considered. What happens to a building in a hot and arid climate, during chilly nights and hot days or sizzling summers and freezing winters? When the sun hits an adobe wall or roof, it takes a long time for the heat to penetrate from outside to inside. This is because the earthen material is a poor conductor and the walls are thick, and because there are sun and shade zones on the curved roofs. By the time the heat enters the interior of the building it is already evening, cooler, and thus the extra heat is not objectionable.

Similar action in reverse happens at night when the interior heat tries to move from inside to outside. By the time the heat goes through the mass to the outside, it is morning and the sun heats the wall from outside and reverses the action. This is one of several cycles of cold and heat transfer during the year, which have similar advantages for the interior spaces. One of the main problems is very hot and long summer days that create unwanted heat inside during the evenings. Traditional solutions have included the use of sunken courtyards, wind catchers, water for evaporative cooling, or simply sleeping outside in the courtyard or on the roof.

In recent years, great amounts of valuable knowledge have been gained by scientific research and data collecting, such as that done on the thermal mass studies of earth walls, which in some ways disproves the traditional beliefs of the great value of thick earth walls; but the danger of using specific knowledge to challenge the validity of earth architecture is to miss the forest for the trees.

2. *Correct Orientation*. Orientation of a building and a community according to the sun and wind is extremely important and should be done with sensitivity to the local and regional climatic conditions. Depending on the climatic or cultural conditions, the living area is sometimes south-facing, toward the sun; but most often on the north side, with its constant shade and cooler air. In the south-oriented buildings, usually an *īvan*, a porch or portico, an overhang, and deepset windows protect the interior spaces from the summer sun and its vertical rays, while it enjoys the direct sunshine in winter when the sun rays are in lower angle. The western orientation is almost always avoided because of the long exposure to the hot afternoon sun.

Orientation according to the wind involves similar basic and perma-

2.1 Borujerdi residence in the desert town of Kashan, Iran. This house has some of the most sophisticated examples of wind catchers, skylights, curved roofs, and interior finishes. The forms are all functional as well as sculptural, and the roofing is a natural mud-straw plaster finish.

nent considerations. Gusty and hot winds must be avoided; the building should be oriented with sensitivity to seasonal winds; openings should be protected; and the vault roof should be oriented for maximum winter sun exposure.

Of course, all this may seem very complicated; and it is sometimes impossible to achieve both sun and wind advantages. But the solution may be much simpler than what we imagine. By observing the local tradition in orientation in an already existing community, and by talking to a few people who have lived in the area for a long time, we can get almost all the information we require.

A house with a good view may be important to some individuals, especially Westerners who are more "extrovert" in their way of life, in contrast to the "introvert" spirit of Eastern societies. Then the tradeoff is a matter of personal preference and the price may be the sacrifice of some of the basic climatic orientation rules. However, if a building is

2.2 Contrasting with the mud-straw roof, the interior of the Borujerdi house integrates finely crafted gypsum plaster finish with the structural and decorative ceiling ribs, wind catchers, and skylights. Such contrasting exteriors and interiors are common in Iran's earth architecture.

meant to be efficient in energy-conscious design then it must be oriented for the sun and wind first. When planning a community, of course, such orientation is vital.

3. *Minimum Exposure*. Minimum exposure to the outside environment, another basic rule, works in close relation with the other rules. Earth architecture, with its inherent unity of the physical and spiritual

characteristics, dictates its own conditions. An adobe unit is small; thus it must be utilized in the form of arch, vault, or dome to cover a space. Because vaults and domes create horizontal loads, they need to be close to each other to balance one another. Thus common walls are essential; thus compact design is created; thus minimum wall surface is exposed to the outside; thus the least amount of heat and cold has the chance to enter.

Such a harmonious relationship exists in the material, form, pattern, color, texture, and spaces created by earth architecture. The pattern of the courtyard houses has been evolved by employing the maximum in structure, the minimum of material, and the ultimate in climatic relationship. Then it is obvious that a house or a community tightly woven together is protected from the harsh environment; it is an organic pattern, just as a tree or a flower is an organic pattern.

4. *Windows and Openings*. Appropriate size and orientation of windows is an important factor in keeping a good balance between the outdoor and indoor atmospheres. Window sizes should also follow traditional as well as regional considerations. The intensity of the desert sunlight allows a small window to bring in ample light for a room. Like an incandescent lamp, the intensity of its light depends on its wattage and not on its size; thus a window should be proportioned to the outdoor "wattage."

Adobe arches are themselves good limiting structural elements that create smaller openings than wood or steel lintels. "Modern" architecture has done great damage to traditional architecture through its use of large windows and giant-size glass panels, considered symbols of modernity throughout the world. In many cases, they have destroyed the appropriate relationship between window size and room space. Today, most Third World windows, traditionally deep-set and small, are being replaced by the large, Western-size windows—even as the Western world changes its codes to restrict window size.

Other factors relating to windows and similar openings include avoiding the afternoon sun; designing window locations across from each other for cross ventilation; and locating the openings away from indirect radiation (the reflected radiation from the ground or neighboring buildings).

5. *Shade*. Finally, we must remember to use different design elements and appropriate technology to create shade, to catch wind and coolness through water evaporation, and so forth. In humid conditions, ample cross ventilation and exhaust is a must. Some design elements—especially the courtyard, wind catcher, and curved roof—are invaluable when incorporated into earth architecture. Porches and shade-giving walls and landscapes also can play an important role in design.

Courtyards

Building around a courtyard, as has been done for centuries in many parts of the world, is a design pattern that has evolved organically. A group of people sits and talks with faces toward each other and backs to the outside; nomads pitch their tents close to each other and protected from the outside—these are such natural patterns.

Such instinctively created patterns can also be observed in animals and insects. The formation of a group of termites against ants is a good example. The termite's very soft lower body can be torn apart easily, but its head and the protecting prongs are its strongest defence. When a group of ants attacks, termites immediately form a circular pattern in which their heads are looking out and their soft bodies are inside. This way they fend with their heads and prongs from the outside, while their bodies are protected on the inside.

The courtyard pattern is an instinctive and logical human solution to living in harsh climates. There are several advantages to this pattern. One advantage is the formation of the minimum number of surfaces against the wind, sun, or outside intruders. If the same number of rooms that are built around a courtyard were built freely, in an open plan, more surfaces would be exposed to the outside and thus less would be protected. Second, the close compaction of the rooms around a courtyard is structurally stronger because they buttress one another; and they are economically more valid because common walls are shared and thus less building materials are needed. Third, building around a courtyard allows the utilization of sun and shade for summer and winter seasons.

The courtyard space itself behaves like a giant chimney to pull up the warm air, while cooler air sinks at lower elevations and into the rooms. If there are basements under the house, and wind catchers, trees, and a pond in the courtyard, a cool microclimate can be created in the hot desert.

In the Middle East a courtyard is an extension of the living room for eating, playing, sleeping, meeting, and praying. It is also developed around the lifestyle of the extended family. As the children get married, more rooms are added around and on top of each other in a courtyard. The concept of the courtyard has also evolved based on the Eastern philosophy of the "introvert." While the extroverted Westerner adores a view and sacrifices much in the way of design to attain that view, many other cultures feel exalted to sleep in a protected courtyard and gaze out at the stars and the dome of heaven at night. This pattern can also be seen as underlying some Native American life patterns, such as the pueblos of New Mexico. These are far removed from the Eastern world, but they share a similar environment. The use of the courtyard as a universal design element is invaluable in earth architecture.

Wind Catchers and Vents

To ventilate and cool a building with wind catchers and vents is another ingenious architectural solution of our ancestors, and one we can learn to use today. Mechanical cooling and air conditioning equipment have done great damage, particularly in rural areas, to the philosophy of a design based on the permanent elements of nature. Had we spent the

2.3 Detail of a sculptured wind catcher and skylight, Borujerdi residence.

last forty years developing a heating or cooling system for our buildings based on the sun or the wind, the whole picture of the building profession in this regard would now be quite different. Our renewed consciousness of the necessity to design with nature, pushed by the high electric bills for coolers and heaters, are all hopeful signs of the forced return to the traditional, basic design elements such as natural ventilation and wind catchers.

The best-known wind catchers (*bâd-gir,* in Persian) used in the Middle East and Africa are those of Iran, Pakistan, and some of the Arab countries. Wind catchers, especially in these arid climates, are very effective when designed to work in conjunction with basements, pools, and fountains. The wind enters a house and goes through the basement or across the pool, thus becoming an evaporative cooling system. Some wind catchers, in Iraq and some Arab countries around the Persian Gulf, are built with a sail-like attachment that can catch the wind from all directions.

A wind catcher can be a simple chimney-like stack, or a structure built on the roof that catches the wind from outside and brings it inside. Air intakes can be arranged behind the wind catcher, taking advantage of the negative air pressure created by unwanted winds to exhaust the warm air from inside.

A wind catcher performs many functions. It can operate as a wind scoop which, on a cool summer night, directly brings the cool breeze inside the building. During the nights when there is no wind, a wind catcher works like a chimney, exhausting the warm air from inside the building. In this case the wind catcher's structural mass releases its gained solar heat to the night sky and causes a draft inside the building. During the days when there is a warm wind, a wind catcher with its inside access doors closed will circulate the wind in and out of its tower, from windward to leeward, and thus create draft inside. When there is no wind during the day, the structural mass of the tower above the roof will lose the coolness it gained the night before and cool the air inside the tower. This cool air sinks down to cool the building. In all these cases the wind catcher works in connection with the windows and openings of the rooms to let the air in or block out the outside atmosphere.

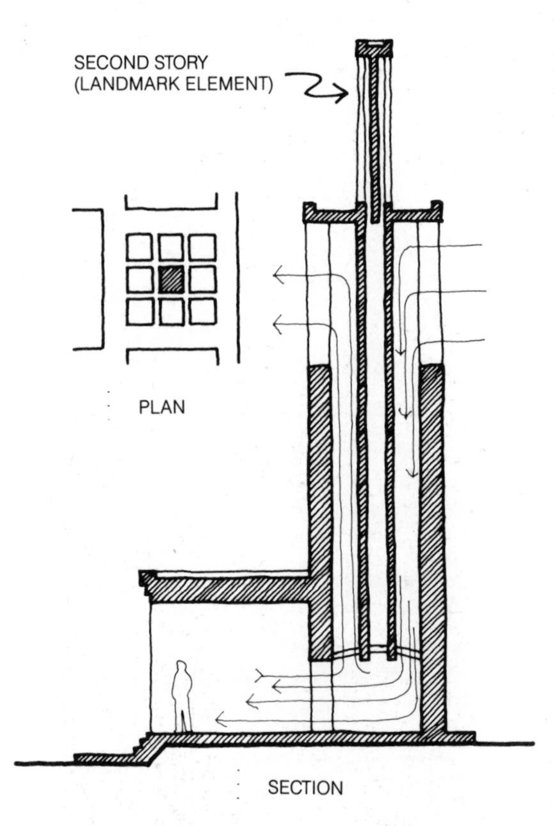

2.4 A two-story wind catcher with nine shafts, Agda, Iran.

A wind catcher works as an evaporative cooler when it is used in conjunction with water. If the wind catcher's access doors are opened in a room with a pond and fountain – especially if the room is a basement – a cooler, more pleasant environment is created. Such basements have

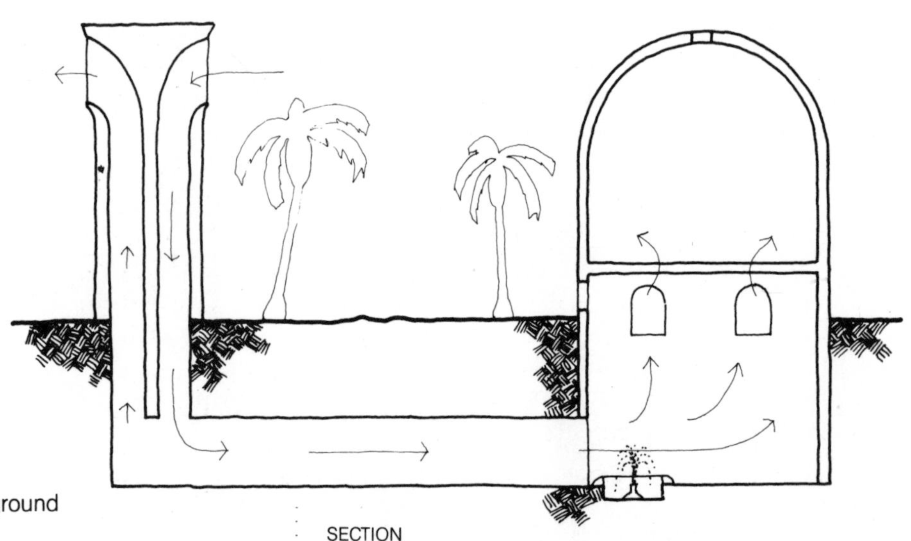

2.5 A wind catcher with an underground tunnel.

2.6 A dome covering an underground water reservoir near the desert town of Yazd, Iran. The two wind catchers keep the water fresh and cool. The sun and shade zones of the dome also create a draft underneath. The dome is covered with fired brick for external protection.

been used as cold-storage houses in some of the desert towns of Iran, such as Yazd, and have been used for living in during the hot summer days as well. Another ingenious use of a wind catcher and water is in the desert town of Bam in Iran, where the wind catcher is built as a separate tower, about a hundred steps away, and is connected to the house by way of an underground tunnel. In the garden and above the tunnel, trees and shrubs are planted. By watering the plants, the tunnel walls are kept damp and the wind passing through the tunnel is cooled by evaporation. In another case, a wind catcher works in conjunction with an underground stream, which usually has cold water and thus cools the warm wind. In many cases wind catchers are built facing in opposite directions: One brings in the wind from one end of the house, and the other exhausts the inside air at the other end. A wind catcher can be built as a simple chimney-like tower; or it can be treated as a sculpture and integrated with the building's aesthetic as a symbolic and functional design element.

Curved Roofs

After the earthquake disasters of Buen Zahra and Tabas in Iran, the local government built some so-called "earthquake proof" concrete and steel houses with flat roofs. But the heat of the summer and the cold of the winter in the cement and steel buildings were unbearable for the occupants. Many of these houses were used by the natives for their animals, while they built their own familiar space with familiar adobe and clay material. And in some of the buildings, where the people had no choice but to accept the shelter, they actually tore down the concrete slab and conventional flat roofs and built an adobe or brick vault and dome, simply because these structures were cooler in the summer.

Functionally, a curved roof works wonders in hot regions. A flat roof absorbs the hot sun all day long, but a dome or a vault breaks the sun into a shade zone and a sun zone. And because the one zone is hotter and one is cooler, an air current is created in the space below. For snow regions, if vaulted roofs are oriented north and south, the sun will melt the snow on both sides of the vault evenly. The curved roof breaks and softens the harsh wind; and, while a wind runs parallel to a flat roof, a dome or a vault touches the wind and creates air currents inside. The curved ceiling is like a collecting funnel, pulling up the hot air; and if a hole or vent is provided at the top of a dome or a vault, the hot air can flow out.

The curved roof surfaces may not be as usable as a flat roof for terraces and sleeping, but the sun and shady valleys between the curves have historically been great places to dry fruits and vegetables. For sleeping purposes, a section of the roof can be flattened by filling the valleys between adjacent vaults with smaller buttressing vaults and sectional walls. This air-filled ductlike structure will act as good insulation against the sun or the cold, and will also help reduce stresses created by the horizontal roof load. A more sophisticated solution, of course, is the double shell dome (discussed in Part 3), which creates greater insulation against the outside temperature.

C·H·A·P·T·E·R 6

STRUCTURAL PRINCIPLES

Building walls with earth and getting help from timber to cover the roof is an ancient method that has been employed successfully all over the world. A bigger challenge to human creativity is earth architecture, which uses the earth alone. The challenge has come about both because of the lack of timber and also the unlimited possibilities of arch-formed structures. The inherent strength and beauty in the form of a curved roof has made its constant appearance possible in history in many ways—rough or cut stone, adobe and masonry, wood panel, concrete, steel or bamboo geodesic domes, and fiber balloons.

Many structural analyses and computer programs are available to explain and help to design the curved forms. But learning some simple natural structural principles can also be a good basis for understanding earth architecture.

STRUCTURE AND NATURE

If we are going to design our structures to be built with earth alone, we need some understanding of the way nature uses the elements to build its structures. As we have said, nature generates structures based on the principle of the minimum material and maximum efficiency. Spider webs, termite hills, leaves, nests, soap bubbles, sea shells, flowers, and molecules all follow this general rule. A spider web is a natural structure that works by ultimate tension, and an eggshell is a structure that works by ultimate compression. Both use the minimum and the appropriate material with maximum efficiency. Just as we learn to build suspension bridges with ropes and cables in imitation of spider webs, we can learn to build domes in imitation of eggshells, building in maximum tension or compression.

When we dump a truckload of earth on the ground, the mound naturally takes a conical shape. When lava flows out of the volcano, it too takes a conical shape; it is dictated by the force of gravity. If we pile up a mound of earth or carve a mountain in a cubical shape, nature will eventually change it back to a cone. The earth dumped by a truck and a volcano formed by nature: both have a similar shape, with a similar angle to the vertical line. This is their angle of repose; they are at rest and in harmony with gravity. When humans discovered how to build a vaulted roof, with leaning arches, their intuition made the connection with earth's gravity, and the angle of repose. If humans with the knowledge of these timeless principles go to the moon or Mars—with different gravity and angle of repose—they can also build arches, vaults, and domes—without the help of form works—in harmony with those planets' gravities.

2.7 The geometric patterns resulting from the natural compaction of circles, which express the inherent unity of circular patterns and have special meaning in Islamic art and architecture.

When we pile up a bunch of equal-diameter pipes or spheres in their natural setting, they all form groups of seven. This is the only way in which nature compacts circular forms. The structural integrity of the triangles created from this natural compaction is a manifestation of the unity inherent in the circular patterns.

If we analyze the soap bubbles formed by circular, triangular, or any conceivable shape frame, we will discover that the bubbles are creating the ultimate spaces that can be formed by the minimum surfaces. Thus nature makes the shortest distance connections with the strongest structure and the minimum material.

And thus we must try to understand, as it is expressed by the smallest particle formations to the largest of the structures of the cosmos, the spirit of unity that exists in the universe. Earth architecture must follow such a spirit.

To understand the natural structures and the equilibrium of the elements and then utilize them in their profound sense is to achieve the ultimate in earth architecture. We can start with an egg. Let us put an egg between the palms of our hands—two ends touching our two palms—and then press as hard as we can. All our power against the thin and fragile egg. The egg will resist an incredible amount of our pressure, and will probably not break even under pressure equal to the load we can lift over our heads. The arch, the vault, and the dome work the same way. If we want to reach such high resistance in our arches, vaults, or domes, then we must try to build the same type of curvature as the egg.

But let us now make a small interpretation of the unity in the seemingly contrasting principle of ultimate tension and ultimate compression in nature. Let us make the egg's curvature from the spider's web. Suppose we have two walls across from each other, and we hang a chain between them. The chain will take the shape of a natural catenary curve. It is in complete tension, which means all the links are pulling each other apart and want to rupture. If we weld all the links together so that they are rigid, and lift the chain upward, we will have an arch that is in complete compression—a catenary or almost a paraboloid arch, as it is called.

Imagine each link in such a chain is an adobe brick, or stone. Each piece presses against the one below; the more gravity pulls, the closer the little cells get together, and the stronger the arch becomes. Working with gravity rather than against gravity—this is being in tune with nature.

Post-and-beam construction does not follow this natural law, because a flat roof is being pulled down by gravity and will eventually fall. Of course, this is no justification for saying that we must always build arched roofs instead of flat roofs, since economic, social, and cultural factors may be of greater concern to us. But such factors and concerns may not be a part of the natural and structural patterns.

Nature makes arches, domes, and vaults; it does not make flat roofs. It gradually changes what is flat into slopes and curves. Building arches, vaults, and domes to the correct, natural, and mathematical proportions will give the strongest possible structures. Today's giant-span arched structures are based on the understanding and calculations for the ultimate in shape and elastic limit of the materials. Thus computer programs allow us to build very large spans with steel, concrete, and cables.

However, less accurate curvature and shallower vaults and domes can safely span more modest distances. It is not only modern materials and correct curvature that will build us safe buildings. Even though today very large spans could be built with ordinary masonry material to last for a long time, it may be more appropriate to leave large spans for modern materials and techniques. The main emphasis when considering earth architecture should be put on the bigger and more pressing problems in the world, such as the need for shelters, rather than to prove the validity of adobe and brick for big spans.

History reminds us, of course, that such work can be done with the old materials: The first largest spanned masonry paraboloid vault was built fourteen centuries ago for the palace of a Persian king. The great hall of the Ctesiphone Palace was built with simple brick, without form work, to span over 25 meters (80 feet). It was 34 meters (112 feet) high and lasted for many centuries. The great masonry vaults and domes of

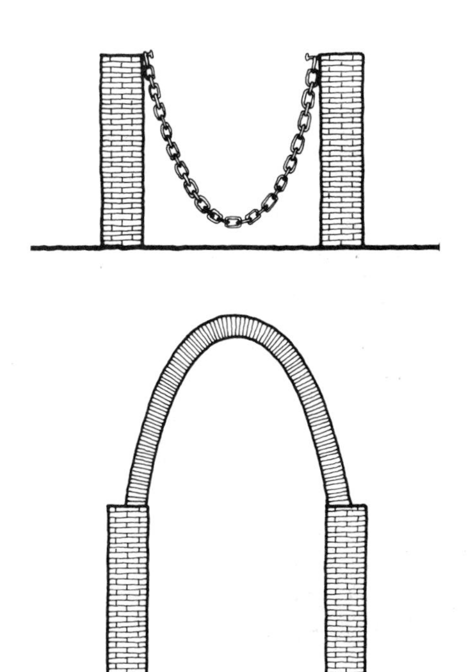

2.8 A chain hanging in a catenary curve (maximum tension) and a catenary arch, vault, or dome (maximum compression).

2.9 The great hall of Ctesiphone Palace (modern day Iraq).

Islamic architecture, as well as Gothic and many other European monumental buildings, successfully were built before any mathematical calculations could prove that they could not be built. Human knowledge and intuition of the natural principles have created great buildings with the simplest materials. Even today in many Middle Eastern cities, such as Yazd and Kashan in Iran, comparatively large spanned domes and vaults built with adobe and mud have lasted for centuries.

Building houses or common-use spaces such as offices, shops, and schools is possible with earth architecture, especially in the appropriate climates. Utilizing arched roofs with shallower ceilings than a catenary curve may be easier, cheaper, and more desirable to build. The simple arch, vault, and dome elements must be built with appropriate spans and heights. The basic concern must not lie in how high the dome or vault must rise or how large a span can be built. The main emphases should go to the fact that buildings, especially shelters, can be created with earth alone, and that everybody can learn how to build one for and with the help of friends or family.

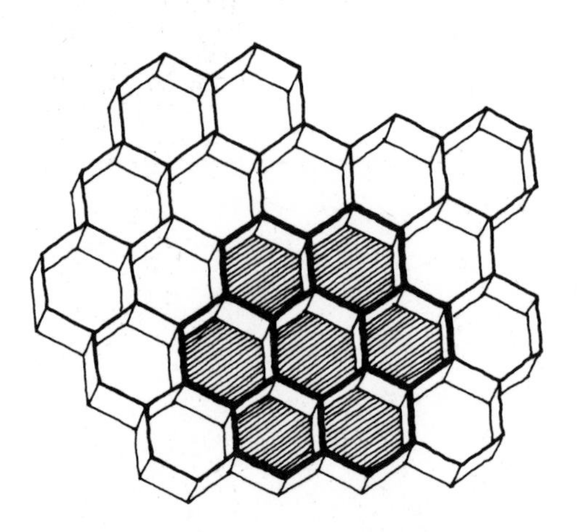

2.10 Honeycomb hexagonal pattern. Maximum spaces are created with minimum material.

Then we must learn how to build with earth alone – no timber, steel, concrete, or plastic – beautifully, safely, economically, and quickly. After we learn to take into consideration all the climatic and socioeconomic factors, we can concentrate on the design, the material, and the method of construction.

HONEYCOMBS, ORANGES, AND ARCHED FINGERS

Let us use the structure of honeycombs and oranges as models for our earth architecture design. A study of the hexagonal pattern of the honeycomb or the radial division of an orange will show the maximum compact plan. It reveals again how nature constructs with minimum material to create maximum space and structure. For real-life design, it means the maximum common walls and minimum exposure to the outside environment. Look again at the honeycomb structure – here is the same hexagonal pattern that is created by the natural compaction of circular units, as discussed earlier.

There are unlimited numbers of such structures in nature, all following the same spirit of unity between material, form, color, pattern, and structure. In earth architecture with flat roofs, we could learn and utilize similar principles; but when we construct with arched roofs, the integration of such principles into the design brings us closer to the natural limits. Using the honeycomb or orange-cut radial plan as a basis for construction with earth walls, arch, vault, and dome forms is a great step towards the use of maximum space, and minimum material. It will create a lasting structure that works with gravity instead of against it.

The arched forms – a simple window, a vaulted or domed roof – work with the force of gravity, rather than against it. Earlier we learned how the form of a catenary arch is in total compression. Now we will look at the equilibrium of these forms when used together.

2.11 A model of an office building, motel, hospital, or market with rows of vaults and a courtyard surrounded by a domed arcade: minimum material, maximum space, short walking distance, short utility lines, minimum exposure, and maximum structural equilibrium. (Rendering by Mr. Pashutan, architect.)

If we make an arch with our thumb and forefinger, and stand this arch over a table top and press on it, we will feel the separation of the two fingers—the legs of the arch—as we press harder and harder on them. To stop the separation of the arched fingers we can form another arch with the other hand and locate it next to the one on the table—like two arches in rows—to counteract the forces. The touching arches will resist all the force we can muster, and will not slip until the fingers are crushed. Similarly, by producing rows of arches, we will have a lasting form as long as the material doesn't crush.

Arches, vaults, and domes work on the same principle. To counteract their forces we must use more arches, vaults, and domes. At the end of the last row or leg, we can either build a very thick wall, or a counteracting structure called a **buttress**. We can also reduce the size of the last arch, vault, or dome to be used as a useful space and at the same time buttress the last leg. The Gothic flying buttresses, the minarets of the mosques, and wind catcher towers are good examples of how the last leg could be stopped from slipping while the buttress itself is used

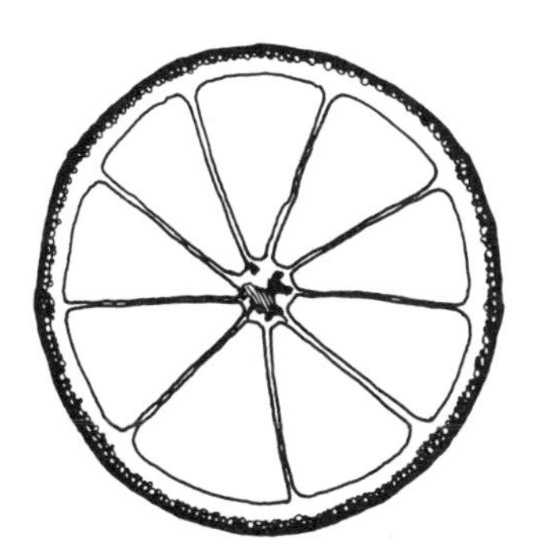

2.12 Radial division of an orange—a compact plan.

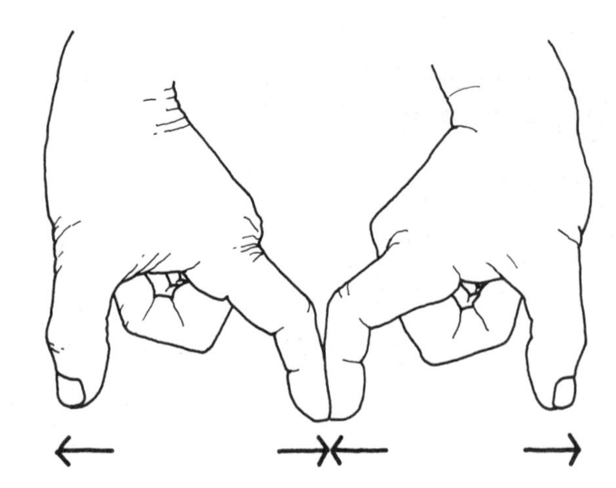

2.13 Counteracting arched fingers.

2.14 Rows of arches balance each other. End arches need buttressing to counter-balance the horizontal force.

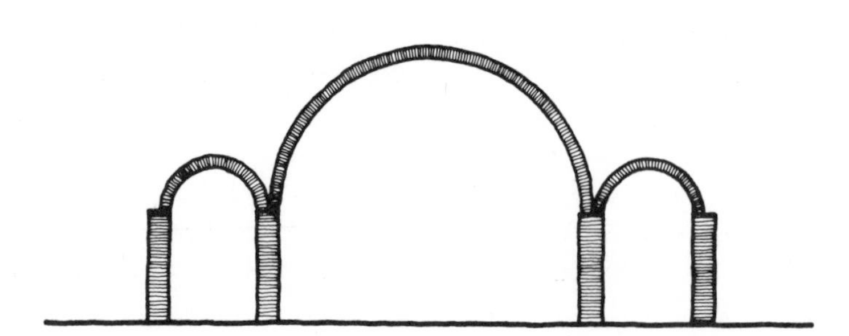

2.15 Small arches buttress the large arch.

architecturally. Then it is obvious that if we build walls with compact patterns—as we have seen in the honeycomb and the orange—with no end wall, and put domes and vaults over them, they will counterbalance each other endlessly and permanently.

C·H·A·P·T·E·R 7

BASIC FORMS AND DESIGNS

If we learn how to build a wall, an arch, a vault, and a dome, we can construct a building with earth as its only material anywhere in the world. But we must only build where it is appropriate. Hot and arid deserts are not the only suitable climates, as is the common belief. Earth buildings have been built all over the world: in the hottest deserts of the Middle East; in the freezing and snowy climate of Scandinavia; and in tropical countries such as Costa Rica in Central America, which has an annual rainfall of 180 cm (70 inches). Thus to build with earth is appropriate everywhere in the world. But it is not the earth alone that makes a building. Skills, availability of the earth itself, local tradition, acceptance, and socioeconomics play a great part in the whole process.

Where appropriate climate and conditions do exist, we could build an entire town with earth alone. Everywhere in the world people know how to construct a wall. Then it is only a matter of learning how to build the other three elements. Structurally speaking, a vault is but a deep arch and a dome a rotating arch. Other curved forms, such as the apse, are simply derivatives of these three basic forms; all stem from the arch.

Here is the rule of thumb for designing a vault or dome roof:

1. A room with a rectangular plan, or close to rectangular shape, should have a vault roof.

2. A square or circular plan room, or a many-sided shape close to a square or circle, should have a dome roof.

3. Rooms that can be broken into semi-rectangular or semi-square shapes should have a combination of vaults and domes. The connecting elements between the spaces are the arches.

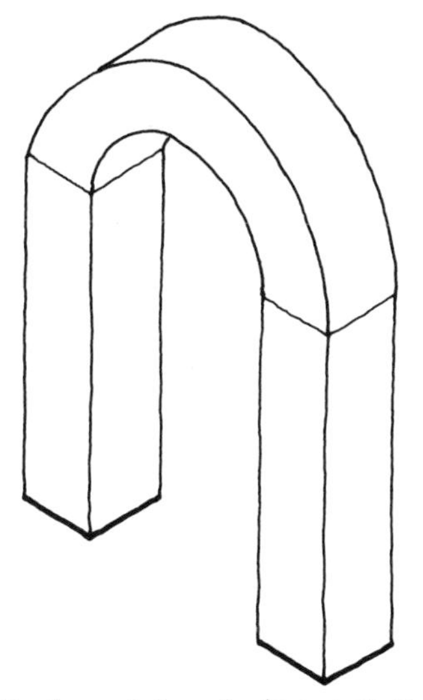

2.16 An arch is a single-curvature structure (curved up and down).

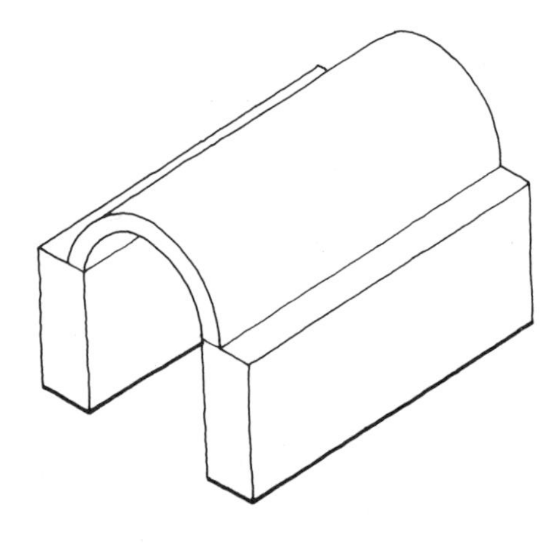

2.17 A vault is a deep arch, a single-curvature structure (many arches stacked back to back).

Construction of single-space or detached rooms should be avoided. A single room with a vault or a dome roof built over the walls will either need very thick walls or end buttresses. If a dome or a vault is built directly on the ground, without walls, or buttressed by the ground, individual spaces can be safely constructed.

Unlimited spatial possibilities are created when the earth is the only material and simple geometry is the pattern. This can be seen in millions of buildings built in the world. To expand even further, beyond the soft earth and the simple geometry the element of fire and human vision could open new frontiers of earth architecture.

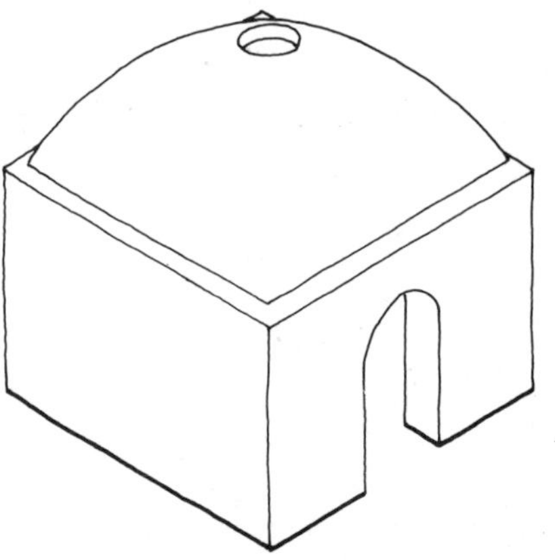

2.18 A dome is a double curvature structure (curved up and down, and sideways).

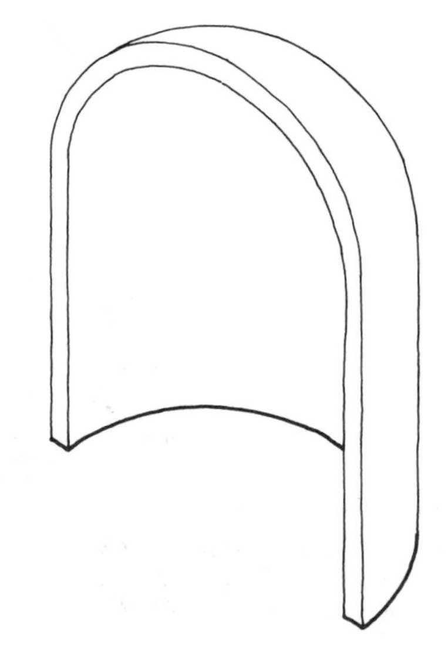

2.19 An apse is a section of a dome (commonly used as an entrance porch).

Earth, air, fire, and
water are obedient creatures,
they are dead to you and me,
but alive at God's presence.

—Rumi

P·A·R·T 3

Materials and Techniques

CHAPTER 8
APPROPRIATE SITE, APPROPRIATE MATERIAL

AN APPROPRIATE SITE

If we are concerned about providing shelters for humans, then an appropriate site is a luxury in which we should not indulge ourselves. It rarely happens that one can afford to purchase an appropriate site for an appropriate earth architecture. We must concentrate on the idea of being fortunate enough to have a piece of land, and we should do our best to build a good building on it. And by a good building we mean one that is strong enough, beautiful enough, efficient enough, and affordable enough. The "enough" is a factor that changes from person to person and condition to condition. It is a grand mistake to try to standardize earth architecture—we have seen the bleak housing projects, faceless office buildings, and other disasters created by the overuse of the International Style of modern architecture.

It is important to learn as much about the site as possible before starting to build on it. We begin by observing where the sun rises and sets and in what direction the winds blow. A design based on the sun and the wind is a good start.

Then we start digging, to learn about the earth. Is the earth good enough to build with, or should we bring other earth to mix with it? Of course, the ideal condition is to dig the soil from our land and construct the building with it. Or dig a basement and use the earth to construct it. But bringing the earth from outside or buying earth-clay is not like buying steel, concrete, or timber. We may be able to dig in a nearby area or purchase it from other spots. The major cost is transportation, which can mean renting a truck or borrowing a few donkeys. Even buying blocks can sometimes be justified, since the earthen material is the only major material we will need.

For those who have the possibility of selecting a site, choose a site with the following attributes for maximum safety and economy: dry land instead of wet and marshy; soil containing clay and sand instead of organic materials and rocks; a site with no chance of being flooded; site with available water, well or piped, and other utilities; good access, size, view, and neighborhood. However, on every piece of land a structure can be built—with an appropriate material and suitable design. Earth architecture has been built on cliffs, marshlands, snowy mountains, and the harshest deserts. This is not to say that earth architecture is suitable to be built anywhere in the world. But since the earth is the most abundant material, with unique environmental qualities, it will find more places in the world to be appropriately used than other materials.

Digging our site and trying to understand the soil and its suitability will require some basic knowledge that anyone can learn. Of course, we could get help from specialists or testing laboratories—but this is a far-fetched idea for most of the world's population. If international organizations or local government bodies could provide people with such basic information as what type of earth is good for what building technique, then leaps in providing shelters would be achieved. But for now, let's learn the way of knowledge for individuals—the homemade tests and ways of finding, extracting, and mixing earth. When we understand the earth, we will be able to use it wisely.

AN APPROPRIATE MATERIAL

Humans are created from clay. These words are inscribed in holy books and the writings of saints, philosophers, and poets. Avicenna, the great Islamic-Persian philosopher and doctor, treated human illness based on the unity of the soul and body and the equilibrium of the elements (earth, water, air, fire) in the human body—the body being the earth-clay and the blood being the fire. In the Persian language, the most common word for "beauty" or "attractiveness" is formed from the word *gel*, meaning clay. *Khosh-gel* means beautiful, good clay, happy clay, and *bad-gel* means ugly, or bad clay. Thus a beautiful woman, flower, or building is molded from a good and happy clay and a bad one is made of a bad clay.

> I wake up thinking about clay.
>
> Why is it that we still know so little about this simple material? Humans have used it since the dawn of civilization, and even earlier.
>
> Ignorance about clay—the substance and its behavior is acknowledged in every book on ceramic, brick, and clay technology. The mystery of clay shows how little we understand the nature under our feet, yet simple men and women making jars can feel their way with clay. Clay is a phenomenon to be understood only through the extension of our senses, not through our logic. The coming of industrialism with its spangles has thoroughly blinded the human creature to the soul that is alive in clay. The protec-

tive layer that industrialized people have developed has numbed their senses.

Concrete and steel have stopped the progress of the clay, which can so beautifully mold imagination into form and space. Vaults and domes, the children of clay, are but imitated in concrete and steel, while the opportunity to create new forms with clay has been stopped somewhere in history. Except for the ceramicist, who is sensitive to earth, today no industry would use the material were it not cheaper than plastic.

Clay is the gift of the Eastern civilization to the West. The altars of the temples of the Western world are made of gold and silver, but here the devout Persian touches his forehead to a bit of clay and bows to God.

There are as many poems in this culture about clay as there are forms in a piece of it. I reach for the book of Omar Khayyam on the shelf and search for his thoughts on the subject. He is most sensitive to clay in the culture of this land. His whole philosophy is kneaded from clay. I find some couplets:

> And strange to tell, among that earthen lot
> Some could articulate, while others not:
> And suddenly one more impatient cried—
> "Who is the Potter, Pray, and who the Pot?"
>
> None answer'd this, but after silence spake
> A vessel of a more ungainly make:
> "They sneer at me for leaning all awry;
> What! did the hand then of the Potter shake?"
>
> Then said another— "Surely not in vain
> My substance from the common earth was ta'en.
> That He who subtly wrought me into shape
> Should stamp me back to common earth again."

. . . One of the hangups that has become almost second nature to us is that we can think of clay only in the scale of our hands. The analogous size of this word, clay, is so limiting that we cannot break away from it. It is like handing a bullet to a primitive man: he can only see and feel the lead in his hands, but has no idea how far it can go. Our understanding of the clay is just that—a vase or a brick that can be moved with our hands, just the way it has been for thousands of years, and no more.

The more we know about clay's ingredients and its technical properties of elasticity, the more we replace it

> with plastics. This is a problem of technology, which gives us only piecemeal knowledge of each individual substance in a test tube.
>
> We can't see the clay as a mass with a soul that might create new phenomena should we, for example, set fire to it. We talk about its low strength and unpredictable behavior. We don't talk about how rocks and mountains are created from it.
>
> The simple elements of water, earth, air, and fire can still create, if the magic of their intimacy is understood, the most perpetual relationship between matter and spirit. . . . How is it that humans created the most beautiful structures and spaces out of clay units before the technology of today stopped them at the elastic limit of this material? How is it that since the strengths of materials have been known and sophisticated calculations have been developed, no architect or engineer dares to build a mud vault of more than a 3-meter span? And yet, far bigger spans were built hundreds of years ago and still stand in the ancient cities, such as Yazd and Kashan, along the great desert.
>
> Those 6-meter mud vaults and domes of Yazd were not built from mathematical theories, but from the freedom of soul to soar beyond the accepted 3-meter limit. To say that we know a lot more about clay and its properties today than at any previous time in history is no justification for the fact that we have lost the feel for it.

Working with the earth requires patience and hard work. If we learn to work with it instead of against it, we will find the rewards to be both physical and spiritual. What is under our feet is not "dirt," is not "soil." It is better than gold.

C·H·A·P·T·E·R 9

ADOBE

Adobe is a sun-dried construction block made of earth. Its main ingredients are clay-earth and sand. Other materials are sometimes added to this mixture to meet the local conditions and markets.

Adobe blocks are usually rectangular or square, and their dimensions vary in different parts of the world. Iranian master masons have found great flexibility in small sizes, such as 20 × 20 × 5 centimeters and its half-size of 10 × 20 × 5 centimeters (8 × 8 × 2 inches and 4 × 8 × 2 inches). They also make shallow grooves by drawing their fingers across the face of the block, parallel to the edges to give the block better adhesion to the mortar. Nubian (Egyptian) masons have been using adobe sizes of 25 × 15 × 5 centimeters (10 × 6 × 2 inches), especially for roofs, with added straw to make the block lighter. They also draw a diagonal groove across the face of the block for better adhesion to the mortar through suction. Such sizes and dimensions are used all over the world because they are usually easy to handle—small blocks can even be tossed in the air to reach workers on the roofs.

The largest and heaviest blocks are used in the West, specifically in the southwestern United States. These blocks measure 25 × 35 × 10 centimeters (10 × 14 × 4 inches), and they weigh anywhere between 15 to 18 kilos (35 to 40 pounds). Only strong and husky men can handle these blocks—smaller or older men, women and children are out of the picture. These huge blocks were used more frequently during the days when Native Americans were taken as prisoners, and the soldiers and prisoners were used as construction workers.

Shelter that is within the reach of ordinary people should be made to fit the people's scale—young and old, men, and women, and children. The size of the construction block or the method of construction should allow the whole family to participate. The smaller sized adobe blocks and the human relation with the physical and spiritual process of earth architecture shows the reason for the abundance of earth buildings in the East. Unfortunately, even though smaller-size adobes are also being used in the West, very heavy blocks are what people visualize as standard. Since most of the building codes are based on large-size adobe blocks, then the block sizes have already been decided by the building codes.

It is a good practice, anywhere in the world, to try to use blocks that are already familiar to the people. If a new block is produced it should be used in conjunction with the familiar ones. In the United States, 20-centimeter- and 40-centimeter- (8-inch and 16-inch-) thick walls are common dimensions.

It is a good general rule to use one size block throughout the building. This will result in the minimum waste. If two sizes are used, the larger ones must be utilized in the construction of the walls and close-to-ground sections, and the smaller ones used for the roof structures.

The adobe size used for general construction and details in this book is the common Iranian block of 20 × 20 × 5 centimeters (8 × 8 × 2 inches), which weighs about 3 to 4 kilos (7 to 9 pounds) and the half-size of the same block, 10 × 20 × 5 centimeters (4 × 8 × 2 inches), which is also a standard fired-brick size in most parts of the world. In addition to its human scale, the small adobe has other advantages. The single wooden form used to make it can be washed in a standard-size water bucket. This becomes particularly important when a high-clay-content mixture is used, since clay sticks to the form. Another advantage to smaller blocks is their flexibility in the wall and roof structure; the ability to toss them in the air for higher elevation use; and the relatively small loss when a block is broken.

Adobe construction is a low-cost, labor-intensive system. It is of great value to almost all Third World countries, because of the availability of cheap and relatively unskilled labor. But this applies only if the whole process is done by hand and with traditional methods. Today we can use machines to make our adobes, use pallets to carry them to the buildings, and use forklifts to bring the blocks within arm's reach of the mason. Even though making adobes by hand is a slow process, and constructing the buildings with hard labor and high spirits has the greatest spiritual rewards, there are other ways of making buildings out of earth.

It is good to determine right from the beginning our own involvement as a builder or user of an adobe building. Earth is the most suitable all-around construction material in most of the world. Getting to know the material and its vast possibility is a good start. And if we are concerned only about the economics, then new tools and methods should be studied and used. For example, to use only big blocks as the quickest route may be a complete misconception. A better way is to calculate the amount of energy, humans, and fuel that will be used to achieve the set target. We may find that it is a better solution to use rammed earth and mud-pile construction (described later in this book). Or we even may want to try a completely new approach. The material is the earth, but the limit is the sky.

THE COMMON ADOBE MIXTURE

Earth, soil, clay, mud, silt, fine earth, sand, straw, and stabilizing material are the most common words we hear when someone is talking about making adobe blocks. The word earth or soil is used to mean the natural, common material under our feet. Clay is the good and rich earth, which we will talk about in greater length. Clay can also mean a mixture of water and earth, the same as mud. Silt, fine earth, or very fine sand may also mean the same thing, referring to very small particles that are coarser than clay, but smaller than sand (just as rocks are broken to grain-size particles as seen in beaches, deserts, and rivers). To simplify the mixture we will talk about clay and sand, fine silt being a part of clayey earth. (Beach sands are not appropriate because of their salt content.) Straw and stabilizing materials are added to the adobe mixture for specific reasons, which will be discussed.

"Clay" is the most important ingredient in every mixture. Clay in adobe is like cement in concrete block, the adhesive element. Even the building codes* usually specify only the minimum or the maximum amount of silt and clay allowed in the mixture of adobe blocks. Then to learn about clay, it could be said, is to learn about adobe.

Let's divide the adobe mixtures into two types. First common adobe, and second fired-structure (*geltaftan*) adobe. Clay will be discussed under the second category.

Generally speaking, almost anywhere in the world we can dig the earth under our feet and build adobe blocks with it. In most cases, the common earth is suitable for our purpose. This is to say that there is enough clay and sand (coarse and fine–sand and silt) in the common earth to make a good mixture. And, of course, a percentage of the block will be a mixture of stones, roots, and other natural materials, which are also acceptable.

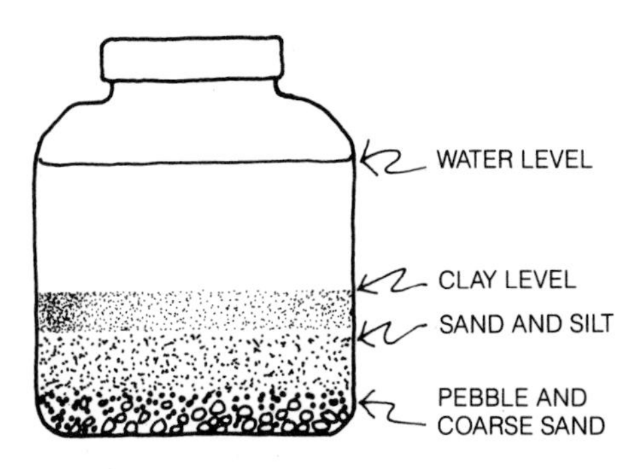

3.1 Soil test jar.

We can learn what type of earth we have by performing a very simple test, the jar test. Pour a handful of earth into a glass jar full of water. Then shake the mixture and observe the settlement of the earth at the bottom of the jar. The lowest layers are the rocks and pebbles; next the sand and silt; and at the highest level, the clay. (Because the clay mixes with the water, it will take a while to settle down and will be the last to separate.) A look at the different layers will give us a good idea about the composition of our earth. From this simple test we can estimate the approximate percentages of each material.

The U.S. adobe building codes specify the amount of silt-clay for adobe blocks to be not less than 25 percent and not more than 45 percent. We have worked with mixtures of much higher percentage very satisfactorily in small blocks.

The best way to test our mixture is also the simplest. The simple test is to make several full-size adobe blocks and let them dry. If there is too much clay in the mixture, the blocks will have many cracks. If there is too much sand, they will break easily and fall apart in the water quickly.

**See the Appendix for a more detailed look at codes, tests, and minimum allowances for adobe mixtures in the United States.*

The earth we use to make our sample blocks must be dug from below the surface, at least as deep as the depth of our hand to the wrist, and it must be free from topsoil and organic materials. When dried, the sample blocks will probably show some cracks and warpage. Cracks are not bad if they are not too large and too many. Some codes allow up to two or three cracks 7.5 centimeters (3 inches) long and 3.5 millimeters (1/8 inch) wide. To see how bad the cracks are, we try to break them either by twisting the block in our hands or dropping the blocks to the ground from above knee height. After making a few hundred blocks, we will be able to tell if the block is right just by looking at it.

The earth and water can be mixed in a pit, in a mechanical mixer, or simply on the ground. An experienced adobe maker in Iran separates a portion of the piled earth, fills it with water, and lets it sit over night to "ripen." The next day he mixes gradually and makes his blocks; he gives the water a chance to seep through as long as possible.

When it is ready, the mixture of clay-sand-water should have the con-

3.2 A simple mixing technique for making clay adobe bricks is performed right at the site.

sistency of bread dough; it should stay in the shovel rather than spill over. Neither the water nor the soil mixture should contain much salt or lime. The only way we can be assured of having the right mixture, the right consistency and the right blocks is to work with the earth, to work with the earth, and to work with the earth.

STRAW

The use of straw in the mixture is practiced in some parts of the world. The straw is used for several reasons. First, it works as a reinforcing element to stop the adobe block or mud plaster from cracking during the drying and shrinkage period. Second, it is a good insulating material and slows the process of heat transfer through the adobe block or mud plaster. Third, it is water repellent and can help the waterproofing ability of the adobe block or mud plaster.

Straw can also be used to make the block lighter or to create special textures. When clay-earth and very fine straw are mixed and "mellowed" by time, they can be used to make a beautiful buff-colored plaster for interior or exterior surfaces. Such natural color plasters, in combination with white plaster trims and edges, have lasted in the Middle East, especially Iran, for centuries.

The amount of straw to use is determined by the type of soil. Making several blocks with various quantities of straw is the best way to find a suitable proportion. A good start may be to use one or more handfuls of straw for five small adobe bricks. The amount of straw used for plastering will depend on use and the exposure of the mud-straw plaster to the elements. The more straw we use, the less surface cracking or erosion.

The straw size should be small, about a finger length for block use and much smaller for plaster use. Large-size straw in exposed surfaces may attract insect nesting.

The use of straw in adobe blocks is not necessary. As a general rule, any extra mixture with the natural earth should be avoided. The basic idea in building with the earth is to use the earth alone. The use of straw is a must for appropriate earth plaster material, but is not necessary for the structural elements.

TREATED ADOBE MIXTURE

"Treated adobe" or "stabilized adobe" are common terms for adobe blocks that have special additives. These additives are used mainly to limit the water absorption of the block. Such mixtures are basically similar to the common blocks, except for addition of a percentage of asphalt emulsion.

Portland cement or other additives are also used to achieve better waterproofing. Stabilized materials must be used for mortar as well. Stabilized or treated adobe is usually manufactured and sold ready-made. Such blocks, if made to correct specifications, can have a long-lasting life without the protection of a stucco-coating.

C·H·A·P·T·E·R 10

GELTAFTAN (FIRED-EARTH) ADOBE MIXTURE

Adobe mixture for *geltaftan*—"fired-earth" or "fired-structure"—differs from a common adobe mixture in its higher clay content and the purity of the mix. The basic ingredients for a *geltaftan* adobe are clay and sand. As a matter of fact, we use as much clay as possible, and enough sand to create fusion in fire. To put it simply: For the best *geltaftan* structure, use the mixture that makes the best fired brick. That means a clean mixture of clay-sand without rocks and organic material. We don't need to use pure clay, as is used in pottery; we can use clay-earth with very small percentages of lime, salt, or organic matter. The mixture for *geltaftan* adobe is similar to the common adobe mixture as far as the method of mixing, time of setting, and method of casting and drying; the main difference is the admixture.

Adobe containing rocks, especially the lime type, will crack after being fired and after being in contact with moisture. Water will cause the fired pieces of lime to expand and crack, or will chip the block. Some impurities can be tolerated, since we are making bricks, not ceramic bowls. After firing and cooling the experienced mason, ceramist, or kiln operator immediately soaks the pieces with water or sprays water over the surfaces, so that the existing lime will get wet and break out. This is done to prevent damage to the plastering. If the interior is plastered, then the plaster moisture will cause the bricks to chip and crack and ruin the plaster work. The adobe made for firing must have rough surfaces. This is usually made with grooves created by dragging fingers through the soft adobe before it comes out of the form.

With the high percentage of clay content in the fire-structure mixture it becomes obvious that we must search for more clay. Let's discuss clay in more detail. What is clay? Where can it be found? And how can it be used?

CLAY

Clay is the fine earth. To the common person, clay is the sticky earth. Our planet is making clay at this moment, just the way it has always been. When most mountains are washed down by rain, the rocks are eventually turned into fine particles of clay. Clay is a type of earth that covers the crust of this planet. Most of the crust of our planet has the chemical substance that is the clay body. In most countries of the world large clay deposits are shown in geological survey maps, but those are only regional deposits. Clay pockets are everywhere and can easily be found by anyone who has a little knowledge and a sensitivity for the earth.

Here is how we can detect clay. Drive a car through a piece of land, or an unpaved road. The more dust that is raised and floats in the air behind the car, the more clay is in the earth. Pick up a handful of the earth and mix it with water. The muddier the water gets, the more clay there is in the soil. There are several lab tests that will tell us all the properties of the earth-clay we have found, but that is too involved to be encouraging.

One of the best ways I have found to locate clay anywhere in the world is to ask the natives, the potters, the brick makers, the farmers, or anyone familiar with the area. They all know where the "good earth," as they usually call it, is located. And if we see any brick buildings around, we know that there is probably lots of clay, because fired brick is heavy and is not usually transported over long distances.

It is simple and helpful to make some field tests to learn about the type of clay we have found. To see how plastic the clay is, we wet the clay to a dough-like consistency and roll it into a coil between the palms of our hands. If the soil has too little clay, it will fall apart. If the earth-clay makes a good, rope-like coil, then try to bend it like a rope. The cracks created at the bend will show us the amount of clay in the earth. Earth that is almost all clay will bend and fold like a piece of rope, with hardly any cracks.

3.3 Clay bar made in the field for shrinkage test.

To find the amount of shrinkage in the earth-clay, we make some flat test bars about 3 centimeters wide by 15 centimeters long and 1½ centimeters thick (1 × 6 × ½ inches). Put this stick-like bar on a table and draw two very thin lines across the bar, exactly 10 centimeters, which is equal to 100 millimeters (4 inches divided into 100 fractions—engineer scale), apart. The reason for making it in centimeters and millimeters, or the 100th fractions of an inch, is for ease of calculation of the percent of shrinkage.

After the bar has dried completely, measure the same distance again. If it is 9.5 centimeters (or 95 fractions, instead of 100), for example, we will know that it has shrunk 5 percent. Some clays shrink a lot, and may need more sand added to the mixture. Common earth-clay shrinks anywhere between 8 percent to 20 percent or even more. If we fire the bar, it will give us the firing shrinkage. The shrinkage due to firing is much less than the shrinkage due to drying, generally from 2 percent to 6 percent.

Since a large amount of clay is needed for the mixture, and since clay

3.4 Scored clay after drying in separating pit.

deposits are not always pure, we should separate the impurities from the clay-earth. If our earth is clay, but all lumpy, then we soak it overnight in a shallow pit and use the mix the next day. (The lumpy clay does not fall apart in the water quickly; it needs time.) If our earth is a mixture of fine clay and pieces of rocks and pebbles, we just use a screen to sift the fine clay-earth.

The best way to extract clay from soil that is mixed—filled with rocks, gravel, and other impurities—is as follows: Dig two pits 30 centimeters (12 inches) apart of around 3 × 3 × 0.3 meters (10 × 10 feet × 12 inches) in gently sloping ground. (If you have flat ground, dig one pit deeper.) Connect them by a spillway-trough. Fill one pit with water, closing the spillway-trough. Mix the soil in the water with a shovel, and let the muddy water spill over to the next pit. Let the clay in the muddy water set in the second pit. Repeat the action until the second pit is filled with muddy water, and let the muddy water settle. At the bottom of the second pit will be clay. All rocks, gravel, and the like will settle in the bottom of the first pit. These must be removed before the next soil-water mix operation is started. (This action could also be done with empty containers in a semi-mechanical way.) This simple operation will give us all the clay we need for mixture to make our adobe blocks, and we can even make pottery out of it.

3.5 Cross section of clay-separating pits.

The best final test is to mix the soil and make several sample adobe blocks. Let them dry out and then fire them at about 1,000°C (1,830°F, or cone 06) in a potter's kiln or in a nearby school kiln. Taking the fired brick and trying to break it will give us an easy answer. Good clay-earth makes strong blocks after firing. If we bang a couple of them against each other, the sound will tell us how good our fired brick is. The sound made by hitting a couple of pieces of unfired or low-fired bricks against each other is a dead sound; well-fired, good clay brick rings like a bell.

We can also test the brick by soaking it with water. Good fired clay brick does not erode or fall apart, no matter how long it soaks in water. An acceptable brick absorbs anywhere from 18 percent to 25 percent of its weight in water. Codes specify the amount.

The firing temperature is around 1,000°C (1,830°F). This temperature is known as "cone 06" in pottery language.

We must remember that we are not making a piece of pottery, we are making a fired brick. Our clay does not have to be as fine and pure as the clay used for pots. It can tolerate some impurities, organic material, salt, lime, or other substances. Ceramic pots made of such clay will crack and fall to pieces; but in a heavy piece of brick, it does not matter as much. As long as it is fired well, won't crack, and can fuse together, some impurities are acceptable. If we look at a good structural, factory-fired brick, we will see that it is dense and has few impurities; but there are other grades of bricks that have lasted in buildings for centuries – handmade, less pure, and less dense. A good rule of thumb is take out all rocks and other organic materials to keep the mixture as soft as clay. Pieces of rock or gravel will fall apart when the block is fired or when it comes in contact with wet mortar or moisture.

Materials for common adobe or *geltaftan* adobe may both be bought and brought to the site. It may seem expensive at first glance; but it is the only material to be transported, not one of several. The proportion of clay and sand does not vary greatly. The clay-earth, almost all clay and silt, could be dug and, without any other mixtures, be made into adobes. The amount of shrinkage resulting in cracking could be controlled by the addition of some fine sands. Large adobe blocks will shrink and crack and will need a lot more sand than small blocks. This is one of the basic reasons for standard, small sizes of fired brick. We can easily use 20 × 20 × 5-centimeter- (8 × 8 × 2-inch-) blocks, which are small enough to avoid shrinkage and cracking. The amount of sand, up to 20 percent or 30 percent, depends on the clay types. But the best mixture could be found by making and firing test blocks and then trying to break the sample bricks. The sophisticated systems of fired-brick industries are unlimited sources of research, of course. But basically the clay has kept its mysteries and humans have yet to completely understand this substance.

CHAPTER 11

ADOBE: FORMS AND BLOCK-MAKING

FORMS

To make adobe blocks of full size and half size, square and rectangular, we have to make forms. The forms are usually made of good quality, smooth wood. Single, double, or multiple forms may be made, depending on what is required. If we are after large and fast production of adobe blocks, and if we have manpower available, then multiple forms can be used. But if the adobes are made by the members of a family, including women and children, then single forms and up to three-section forms are practical for full-size blocks. For half-size blocks, two- to six-section forms can be used.

For full blocks, the forms are 20 × 20 × 5 centimeters (8 × 8 × 2 inches); and half-size forms are 10 × 20 × 5 centimeters (4 × 8 × 2 inches). When making the forms, the sharp edges and corners should be sanded smooth to avoid injury. Sometimes metal sheets are used to tie the wooden pieces together for a better, more uniform fit with the least number of nails. Metal sheets may create sharp edges if the forms are not handled with care.

A skilled and motivated adobe maker can make all the adobes required with a single form and a helper, as quickly as the blocks could be made with multiple forms. One adobe maker and his son, using one adobe form for full-size blocks and one five-section form for half-size blocks, made all the blocks needed to construct a ten-classroom school—with only one month of lead time before construction began.

After the adobe is dried out, the block will be smaller than the form. If we want to make a more accurate block size, then we must first learn about our mixture and its shrinkage-percentage and make the form that much larger. However, for general use and non-machine-made blocks, such accuracy is not needed; shrinkage can be accommodated by adjusting the mortar joints during construction.

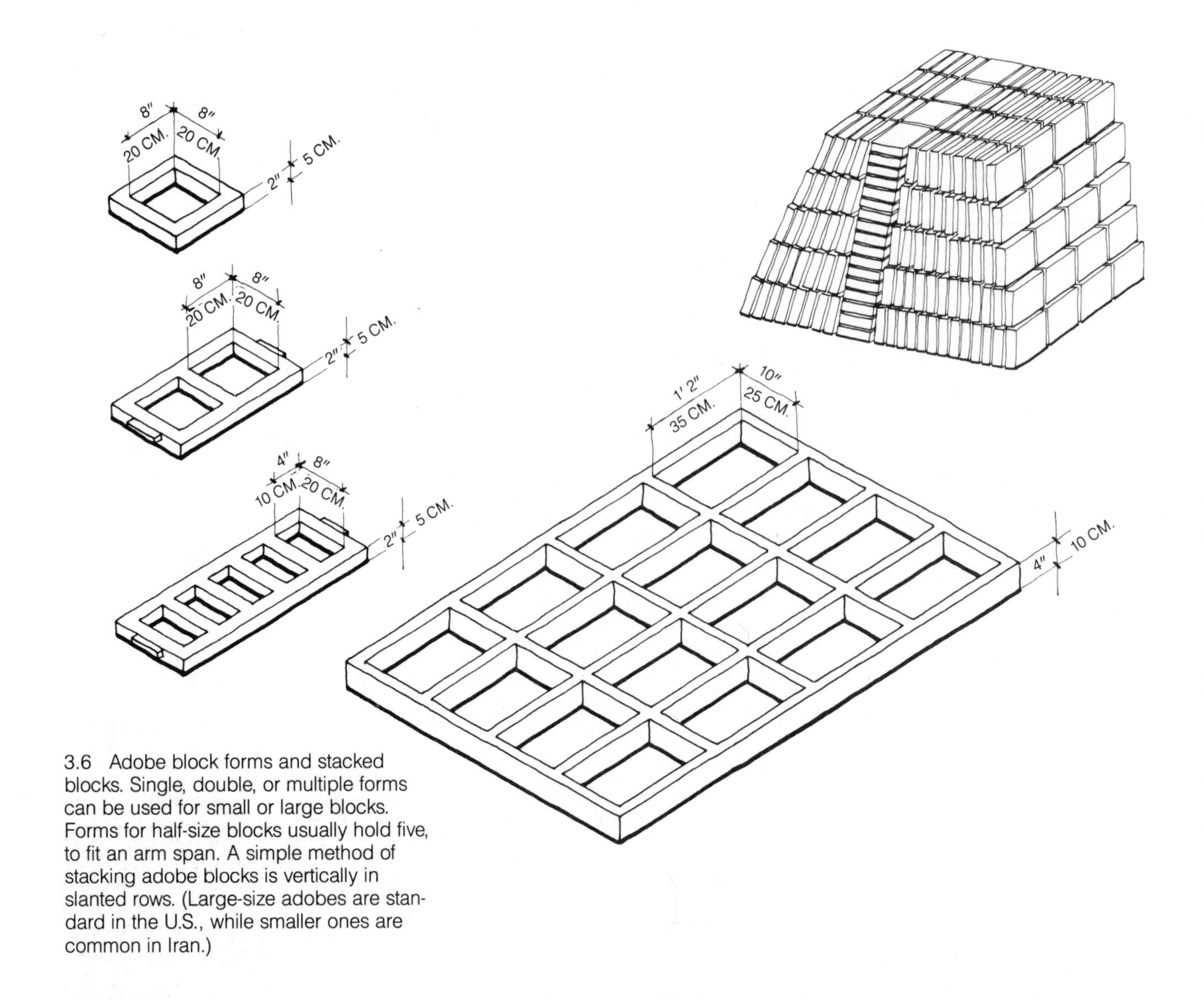

3.6 Adobe block forms and stacked blocks. Single, double, or multiple forms can be used for small or large blocks. Forms for half-size blocks usually hold five, to fit an arm span. A simple method of stacking adobe blocks is vertically in slanted rows. (Large-size adobes are standard in the U.S., while smaller ones are common in Iran.)

BLOCK-MAKING

1. The day before block-making begins, transport enough adobe mixture for a day's work to the block-making area. The closer at hand the mixture is, the faster adobe making will go.

2. Have a thorough knowledge of the local weather conditions. Wet or freezing weather will destroy adobe-making efforts.

3. Designate a piece of flat, raw land for the operation and sprinkle the area with some sand. This will let the block breathe, and will also absorb the extra water during the drying period. Blocks made on an asphalt surface, for example, are subject to nonuniform drying and cracking. As a general rule, to prevent adobe blocks from sticking to the ground, or to prevent wet mixture from sticking to the wheelbarrow, we can use sand, straw, or just some dry adobe mixture.

4. Soak the form in water. The water container, usually a bucket, should be right next to the adobe maker for constant use. It is best to dip the form in the water before making every other block.

3.7 An Iranian adobe maker, with a single form, a bucket of water, and his son assisting him, makes all the blocks for a ten-classroom school.

3.8 A Mexican adobe maker (*adobero*) uses a wheelbarrow, a bucket, and a four-block form.

5. Lay the form on the ground. Pour the earth-clay mixture into the form with a shovel, or with both hands if the material is next to you and your form is small. With both hands, work the mixture well into the corners of the form so that there are no gaps. Then flatten the top of the block even with the form edges by wetting and rubbing with the hands. For the final touch, after the surface is even, make shallow grooves with eight fingers in a parallel line from edge to edge. These grooves will help the block and mortar make a tight bond. When the block is used to build a curved roof, the grooves help reduce the chances of slippage. And for fired-structures, the grooves help the fusion process. Then hold the form by both hands and pull up, and start the next one.

6. Dry the blocks flat for a day or more, depending on the weather. Then stand the blocks on edge to dry evenly. Wait one more day before stacking them to sit for the total drying period, which may take between one and two weeks.

Stack the adobes on the ground as close together as possible, for minimum land use and ease of stacking. Each row of adobe can be made with no separation between the blocks (other than the dividing form thickness), which makes a little distance in between it and the next row. At the start, if the form sticks to the adobe and the block doesn't come out easily, this just means we must practice more.

To check the adobe for dryness after the total drying period, break one and look at its color. A uniform color means a dry adobe; shades of light and dark at the broken edge means there are wetter and dryer parts. During the drying period the adobe blocks will develop some cracks, due to the shrinkage. Most cracked adobes are acceptable if their cracks are

3.9 Mechanized adobe making in the United States, San Juan Pueblo adobe yard, New Mexico.

not too many, too long, and too deep. Dropping a block from about knee-high to the ground will show how bad the cracks are. In some building codes these cracks are specified and their acceptability defined. For example, Western codes usually accept up to three cracks, each one as long as 75 millimeters and as wide as 3 millimeters (3 inches long and ⅛ inch wide). Cracks may be caused by too much clay, too fast drying, or too large block size; but incorrect mixture proportions are the main cause.

After the first few days of drying, the adobes must be stacked on their edges for the one-to two-week drying period. The stacking is done to clear the ground for the next batch, for ease of transportation, and for better protection.

Many patterns of stacking are used in different parts of the world. In brick-manufacturing yards, where the kilns are burning all year round, the large amount of adobe blocks are made during the warm seasons, and saved for the winter. These large masses of tightly stacked earth blocks are sometimes covered, only at the top, with a very thick layer of the same earth to protect the blocks from rain and snow. Adobe blocks, once made and dried, can resist several showers of rain, depending on the amount of clay in the mixture.

MORTAR MIXTURE

The best mortar for common adobe construction is made from the same material as the adobe blocks. The important difference is the amount of rocks and organic materials. The mortar must be as free from these materials as possible. Screened earth will give us a good clay-sand combination for mortar. In the old Persian tradition, to separate the rocks from mortar or even adobe mixes, the owner would drop coins into the mud mixture and the workers or the mason would try to find the coins. They would pick out and throw away the rocks in their efforts to get the money. Such practices bring humor and good results to the work.

Use of additives such as lime, sand, straw, cement, or other stabilizers in the mix is not necessary, and may even be damaging because their reaction to weather will differ from pure-earth adobe blocks (unless, of course, the block is made from the similar materials).

Mortar for fired-structure construction is also the same material as its adobe, except when a better fusion is desired between the block and the mortar. The best materials to use in the mortar mixture are the common ceramic glazing mixtures such as glass, soda, or Colmanite. Colmanite (known among potters as gerstley borate) is a natural mineral used in powder form and mixed with water. It is relatively abundant and inexpensive in some parts of the world, and mixes with clay and sand easily to make a good mortar. Colmanite, soda, and glass are **fluxes** – materials used in ceramics to create better fusion and lower the melting temperature of the mix. Fluxes will help melt and fuse the mortar in the relatively low temperature used for firing structures (around 1,000°C or 1,830°F).

3.10 Machine-made blocks are stacked on pallets and delivered to the site.

The amount of Colmanite or other fluxes in the mortar varies depending on the clay, the mixture, or the fire.

It is best to mix some clay-sand mortar and add one or more of the fluxes in 5 to 20 percent test samples. We make some block samples from the adobe mix we will be using, and stick them together with this mortar, and fire them. We will then see which mixture provides the best bond, and can adjust the mortar mix accordingly. Local potters can be of great help in these determinations. Hundreds of tests have been carried out by my students to find the most available, least expensive, and basic mixture materials around the world. The result showed very promising fusions created by the following mixtures:

- clay and table salt
- clay, lime, borax, and crushed glass
- clay, crushed glass, and slag
- clay and borax
- clay and crushed glass
- clay, sand, and colmanite

CHAPTER 12
FOUNDATIONS

Why do we need a foundation? Any type of ground can resist only a certain amount of load, dead or live, before it cracks, slides, caves in, or slips. Dead loads include the weights of our walls, roofs, and so forth; and live loads include people, furniture, and snow. These are all vertical loads coming to the ground via our structure, and the ground should support them safely.

Earthquake and wind loads are similar to live loads, but they are mostly horizontal–they push, pull, and uplift, creating unbalanced loads both for the building and the ground. There are also pressures and reactions from the ground–moisture, freezing, and swelling can uplift buildings. And adding to all these comes the greatest enemy of all for buildings built with earth: wet weather.

With all these problems surrounding our buildings, if we add social, cultural, and economic problems to them, it seems amazing that humans even dare to pour foundation to their shelters. But building shelters, even more than finding food, has always been foremost in the human mind and soul; and despite all the problems, perhaps the greatest portion of human ingenuity has gone into creating buildings. Humans are builders by nature.

We don't need to worry about many of the factors concerned in building other foundations, since our earth structure is simple and is a part of the earth itself. There are many buildings built with earth-adobe blocks or *chineh* ("mud-pile" in Persian) in the world that do not have a special foundation. The earth is simply dug to a shallow depth and the adobe blocks are set right from the ground, up to and including the roof. Many such buildings in the hot and arid climates of the Middle East have lasted for centuries.

Use of concrete for adobe foundations has become common practice, especially in the Western world, where cement and steel reinforcing is relatively inexpensive and abundant.

While in many parts of the world a bag of Portland cement is a black-market item, in the West it is quite common. Almost all building codes are deeply involved with the use of such manufactured and tested materials. Even for common adobe buildings, it has now become a great challenge to replace the required concrete items in earth buildings with materials other than concrete. The inherent strength and predictability of concrete members has pushed building officials to the easy-out explanations of the charts; earth structures, on the other hand, are not so easy to categorize and thus are harder to explain in official language. The time-tested adobe structures are good evidence to use to challenge the set rules and regulations pushed on earth structures by the codes. The more strict the codes, the less chance there is that we will develop affordable shelters.

The use of manufactured materials such as cement or steel in adobe buildings becomes ridiculous in much Third World construction. Cement, for example, is often mixed with salty water and muddy sands to make concrete. And it is a proven fact that some concrete tests in these areas have shown lower compressive strength than the common adobe. Use of concrete foundation and treated adobe blocks in the West must not be a reason to use modern materials without technology to back them up. The basic indigenous and natural materials of clay, sand, lime, and rocks can still create the most durable foundations, if used with the tradition.

Fortunately, the great wave of interest shown by Westerners in adobe, rammed earth, and other earth architecture has created many challenges to the dogmas of the Western building codes. And it won't be too long before many of the traditional ways and materials will become Western trends as well. And then it would be scornful to see the industrialized West choosing the tradition of Third World construction systems, while the Third World is busy imitating the discarded ways of the Western world.

Before we begin building the foundation to our earth building, we must learn about the earth on our site. It is also a good idea to find out what kind of foundations nearby buildings have, and what the local building traditions are. Even though each site in the area may have a different type of soil, long experience usually leads the whole region to build similar foundations.

The depth of the foundation is mainly determined by the frost depth, since freezing makes the earth swell and lift and crack the walls. For example, if the ground freezes to about 2 feet deep, then the bottom of the foundation should be below that. Earth architecture is commonly found in the desert regions of the world, and there are a lot of freezing desert climates, despite the connotation that desert means "hot." But since there is hardly any ground moisture in arid regions, then even the freezing depth may not be a serious factor. In countries or regions with building codes, the minimum depth for foundations is usually predetermined. Even though their validity is questionable, the codes must be followed.

TYPES OF SOIL

Unless the site and soil is unusual, the following simple rules should be enough to help us build a solid foundation. After digging into a piece of ground to, say, knee-high, we may find one of the following situations:

1. A hard ground with a mixture of rocks, gravel, sand, and the clay that has glued them together. This conglomerate soil will make a solid ground for our foundation.

2. Rocky ground, which is solid and fine to build on.

3. Clay ground must be looked at for suitability. If the soil is solid, dry clay, it is fit to be built on. But if the ground is a wet clay, it is not suitable—especially if the wet clay land is on a slope, which may cause the foundation to slip. The wet clay ground creates slippage and cracks under the load differences of the walls.

4. Organic soil or fill is a weak ground. It should either be avoided, or the foundation should be dug deep enough to rest on the solid natural and dry soil.

The foundation must be built on a piece of land with a uniform type of soil. Differences in soil condition under different walls can cause uneven settling and cracks.

COMMON ADOBE FOUNDATIONS

Once suitable soil is found, we dig a foundation ditch that is 20 to 30 centimeters (8 to 12 inches) wider than our walls—20 centimeters (8 inches) wider if we want to use concrete footing, and 30 centimeters (12 inches) wider if we want to use rock footing. Rock foundations are built with clay, cement, or a lime-clay-sand mortar. The rocks must fit snugly so that if floods and heavy rains wash the mortar away, the foundation won't fall apart. The foundation under bearing walls must be wider than

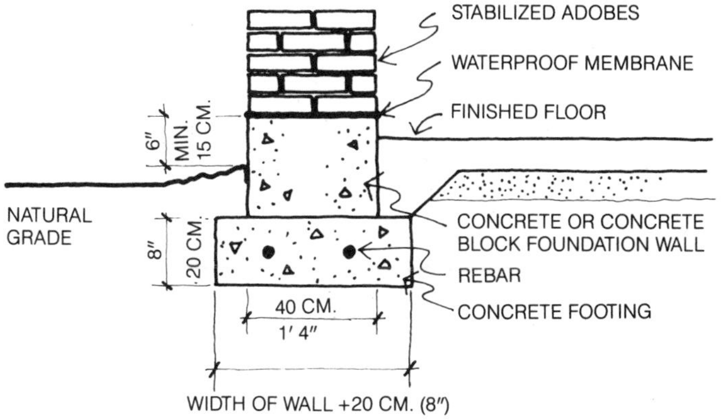

3.11 Concrete foundation.

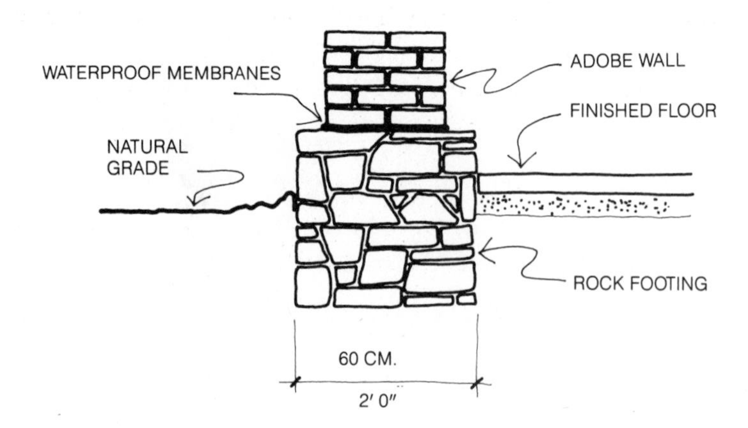

3.12 Rock foundation.

the wall itself to give it more bearing capability. But for partition walls, a wider foundation is not generally needed. Foundations with packed gravel as the only material can also be built on.

A very common material used for foundation in some parts of the world is earth and lime or lime concrete (called *shefteh*, in Persian). *Shefteh* is a mixture of earth-clay, sand, and lime with a proportion of about 250 kilograms (550 pounds) of lime powder to one cubic meter (35 cubic feet) of common adobe soil that has clay and sand in it. Sometimes pieces of rocks are also added to the mix. After the mixture is placed in the foundation ditch, we let it sit overnight to lose its water to the ground. The next day we compact the foundation with a hand ram. For deeper foundations, *shefteh* is poured and rammed in layers of 30 centimeters (12 inches). Lime mixes of different proportions have successfully been used in the Third World and the West as water resistant material.

If the soil is dry clay, and the climate is hot and arid, we can dig our foundation trench and start our adobe wall right from the first layer into the ground. Generally, no rock or concrete foundation or any special treatment is needed in very arid climates. Common adobe must always be protected from water, especially from ground water, which seeps up the wall by capillary action. To have a lasting adobe wall it is advisable to build—even if building codes don't specify—a water resistant foundation wall that extends above the natural grade at least 15 centimeters (6 inches). However, in dry desert areas of the Middle East, walls that have no foundations have been standing for centuries.

Treated adobe may also be used to build the foundation walls. The foundation wall, or the stem, is the short wall built over the footing up to a distance above the natural grade. If the foundation is shallow, then the footing is built as one piece to the above-grade height; but if the foundation is deep, then the footing is built wider and it is continued, with the same material, in a narrower width as the foundation wall to the above-grade elevation.

After the foundation has been built with concrete or rocks to 15 centimeters (6 inches) above the natural grade, the adobe wall can begin. A layer of tar paper or lime-cement mortar waterproofing could be laid in the joint between the top of the foundation wall and the first layer of the adobe, although in normal conditions it is not needed. A Portland cement mortar cap can also be built as waterproofing over the foundation wall. When concrete block foundation wall is used the cavities are filled with rocks, and then capped with a layer of Portland cement mortar. The waterproofing will stop seepage of water from the foundation wall to the adobe wall.

GELTAFTAN FOUNDATIONS

The foundation for fired structures isn't much different from the common adobe foundation, except in some cases. If we want to fire the walls all the way down to below-grade level (that is, including the adobe-block foundation), then we leave the foundation walls exposed to the fire. To achieve a uniform fired-foundation we must provide **flue** openings at the lowest part of the structure, so that the fire is pulled down to the lowest adobes. (Flues are discussed in Part 4.) The base of the foundation is over a layer of volcanic sand-gravel for base isolation.

After the firing is done, fill and pack the trench at the foundation wall with loose gravel and cover it with packed clay. If we want to fire the finish floor at the same time, we leave a distance equivalent to a fired block–say 20 centimeters (8 inches)–between the inner face of the foundation wall to the inside floor line; after firing, we just cover the inside trench with a fired brick to match the finished floor. In this case it may be preferable to fire the structure first, including the foundation, and construct floor slab and finishes later. If flooring is uniformly constructed with the foundation, it must also have a packed base.

If our foundation is conventional rocks or concrete, separate the adobe walls that are to be fired from the foundation wall by four or more layers of pre-fired bricks. This creates a fired-brick buffer zone between the rock

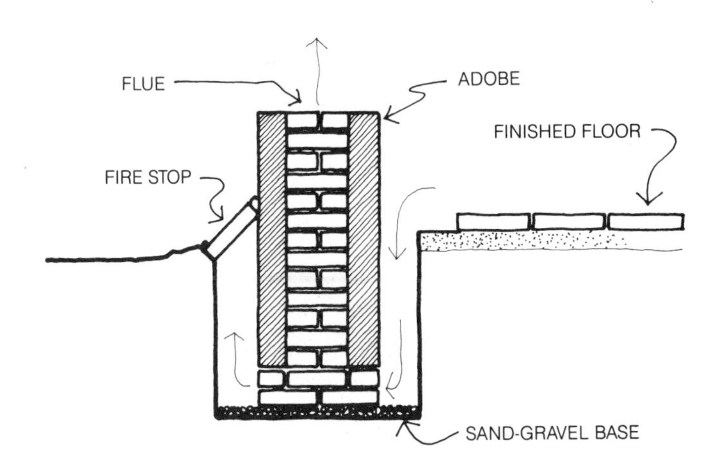

3.13 Uniform *geltaftan* foundation section at a flue.

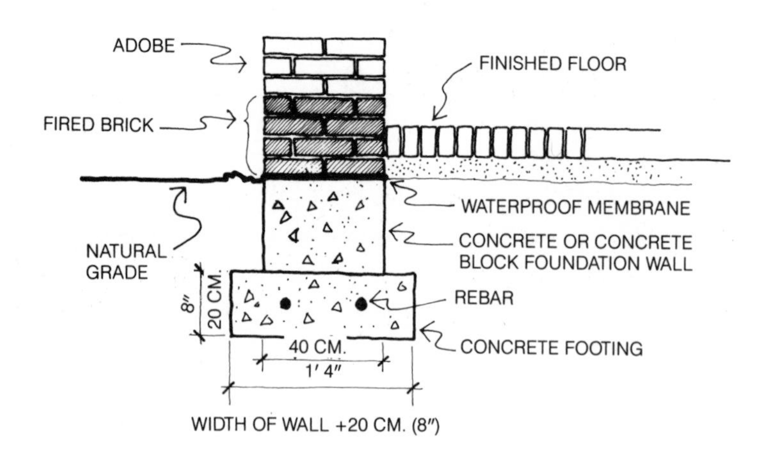

3.14 Standard foundation modified for firing.

or concrete foundation wall and the adobe wall. Even if waterproofing membrane is used, it will be in the lower level, under the pre-fired bricks, where the fire cannot melt or destroy it. In this condition, the floor could also be fired along with the walls, since the flue openings are at the underside level of the finished floor. In the area of the flue openings, the floor-slab adobes could either be left out and filled later; or mortar spaces between the floor blocks could be left open so that the fire can go through the floor and up the flue. Constructing the floor after the firing of the structure can be a simpler job. Generally speaking, the fire could be led into any area, below or above the foundation, depending where the level of the flue openings are located. This is similar to the **downdraft** system of kiln construction (discussed in a later chapter). It means the fire has to bake all the surfaces before it is vented to the outside through the flue lines. The firing is done by the combination method of **updraft** and downdraft systems, discussed in Part 4. To standardize uniform or conventional *geltaftan* foundations and walls, tests in different regions of the world with varieties of earth-clay materials and firing techniques must be carried out. The details explained above are based on experimental design data.

CHAPTER 13

WALLS

Adobe walls for buildings with dome and vault roofs are usually thicker than the walls for flat roofs. This is because we are building more space with more material (of the roof) and also because curved roofs create both horizontal and vertical loads.

The thickness of the walls, then, will depend on roof span, roof height, and the amount of compressive strength our adobe blocks have. For large projects, all these must be calculated and laboratory tested. But for typical, short-span, small buildings—which means most buildings in the world—such technical information is neither available nor necessary. A few experimental rooms built and tested locally will give all the needed information. (Local governments and institutions would do a great service to provide such safety information, rather than restricted codes, which often succeed only in creating great discouragement.)

COMMON ADOBE WALLS

The typical supporting wall for a span of 3 to 4 meters (10 to 14 feet) for a dome or vault is between 40 to 60 centimeters (16 to 24 inches) thick, depending on the height of the wall and whether or not the structure is fired. The general rule is to build a solid wall, with minimum openings, to support the roof. In a vaulted room, which is rectangular in shape, the load of the roof is on two side walls only; thus it is possible to have large windows in the end walls. In a domed room, which is close to a square shape (depending on dome style), four walls and/or the corners carry the roof. In rooms that combine domes and vaults, the supporting arch acts as the supporting wall. Door and window openings and anchoring of their frames must be well thought out before the walls are constructed.

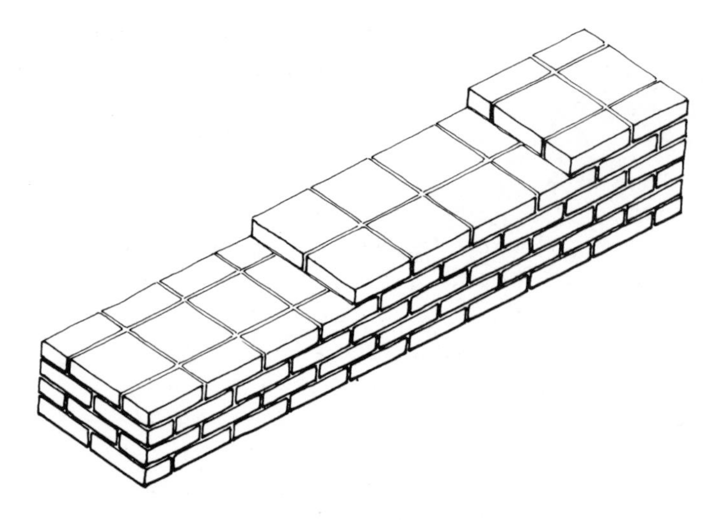

3.15 Wall two adobe blocks wide. Layers of two full blocks alternate with layers of one full and two half blocks. Blocks are laid next to each other without mortar in the vertical joints.

Laying of the walls, depending on the thickness of the wall and the adobe sizes, follows these general rules:

1. Lay the blocks so that all joints are staggered in alternate rows.
2. Build the wall straight.
3. Make sure the wall is solid.
4. Use mortar made of the same ingredients as the adobe.

Walls are built with both square adobes, 20 × 20 × 5 centimeters (8 × 8 × 2 inches), and rectangular adobes one-half size of the square, 20 × 10 × 5 centimeters (8 × 4 × 2 inches). The number of square adobes used in a building is around twice the number of the rectangular ones. (This proportion changes at the roof, which may be built of all square or all rectangular adobes.) One of the simplest wall construction methods is the double and single alternative for two-adobe-thick walls, and a similar system for a three-adobe-thick wall. If larger adobe blocks are used the pattern can differ, but it must follow the general rules. Building a curved wall onto the end of a rectangular room, or wherever a vault starts, is discussed later in Part 3 under vaults. Some of the code requirements for bond beams and horizontal and vertical reinforcing are in the Appendix.

For fired structures, the adobes are put next to each other with *no mortar in the vertical joints*. Each row is laid over the mortar on the row below. The ends of the adobes actually touch each other, dry and without mortar. This type of wall construction is suitable for firing, as well as for a common wall. But if the horizontal and vertical joints were both filled with mortar, the fire would not penetrate the wall easily and there would be no room for expansion and contraction during firing.

Walls for fired structures are usually built two adobes 40 centimeters (16 inches) thick. The wall must continue in a straight line up to the spring line (the top of the wall where the dome or vault starts). The arch openings in the wall for doors and windows must be constructed so that they integrate with the walls to create a bond at the top.

Vertical reinforcement of the adobe walls is dealt with in local codes, which vary as far as numbers, sizes, and distances are concerned. In most parts of the world however, reinforcing bars are neither available nor affordable. There tree trunks, bamboo, adobe-mud-pile, adobe-rock and other combinations are used in common earth buildings.

Many very tall unreinforced adobe or mud-pile walls – either standing alone or as supporting structures – still stand after centuries. But such walls may not be safe or practical in high seismic zones. One-story undulating walls without reinforcing have been successfully used in some earthquake regions. For example, a young Iranian architect-engineer successfully built a housing complex in the great Buena Zahra seismic area, using unreinforced masonry undulating walls over a sand-gravel foundation. Careful tests and calculations were done beforehand, and the results were successful. In spite of its success, however, the local building bureaucrats ignored the project and continued to build their pet precast concrete buildings.

That unreinforced adobe and masonry buildings survived the Tabas earthquake shows the possibilities of combating seismic forces without reinforcing steel. And if researchers would continue to explore options other than only using steel and concrete for earthquake safety, innumerable lives and buildings could be saved in seismic areas in Third World countries, where the poor could never afford manufactured materials.

Partitions and nonbearing walls are commonly built in single-adobe width of 20 centimeters (8 inches). Lightweight walls are constructed with adobes used in hollow box patterns. In this case, each six adobe blocks make a hollow cubic box – side blocks that are standing on edge. Such walls have been used in Iran as parapet walls on the roof. Thin *curatin* walls, 5 centimeters (2 inches) thick, called *tigheh* ("blade-thin") in Persian, are constructed with adobe blocks and laid up on edge to close small openings in walls. Larger *tigheh* walls are commonly constructed with fired brick, laid up on edge and cemented with quick-set gypsum mortar.

RAMMED EARTH WALLS

A vertical slice of solid natural ground is like a rammed earth wall. By piling and tamping damp earth, we can make a solid mass of wall in the shape of its mold. To make a rammed earth wall, follow these four steps:

1. Construct a wooden cavity form for the wall.

2. Place damp earth inside the form, in 20-centimeter- (8-inch-) layers at a time.

3. Ram the earth in consecutive layers with a heavy device.

3.16 Rammed earth walls with different compositions, textures, and orientations built at the Environmental Research Laboratory of University of Arizona, Tucson. Architect Helen J. Kessler has conducted the research.

4. Remove the forms.

The material for the rammed earth is the same as for common adobe block.

Rammed earth walls, called *pisé* in French, are especially promising for industrial countries, where labor is expensive and equipment relatively inexpensive. The technique can be even more successful than the adobe block system, particularly in projects where equipment is available and forms could be used repeatedly. In damp climates, the rammed earth method is faster than adobe block construction, because rammed earth walls do not require the drying time before use that adobe blocks do.

Forms used for constructing rammed earth walls are similar to concrete forms, and can be simple or complex. Ramming devices can range from a wooden pole to a machine-operated pneumatic system. In recent years, much progress has been made both in research and development of this technique, especially in the United States. Firing rammed earth walls is similar to firing mud-pile walls. The rammed earth structure must be fired from both sides.

BUTTRESSES

A buttress wall is a wall built at an angle for shoring and bracing. Buttresses have been used as structural, functional, and sculptural elements.

The common wall between spaces supporting vaults or domes carries the roof loads vertically down to the ground. The end walls must

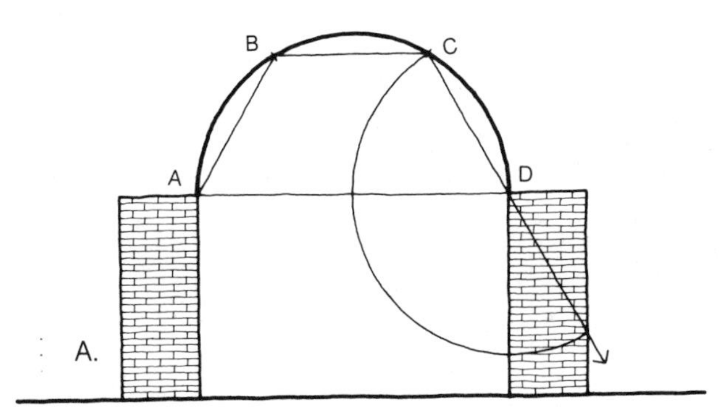

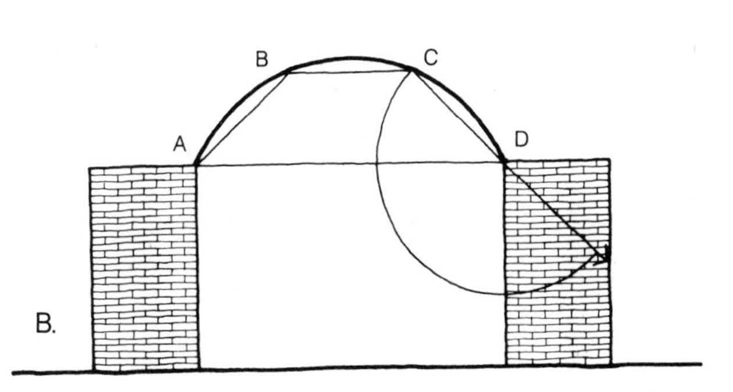

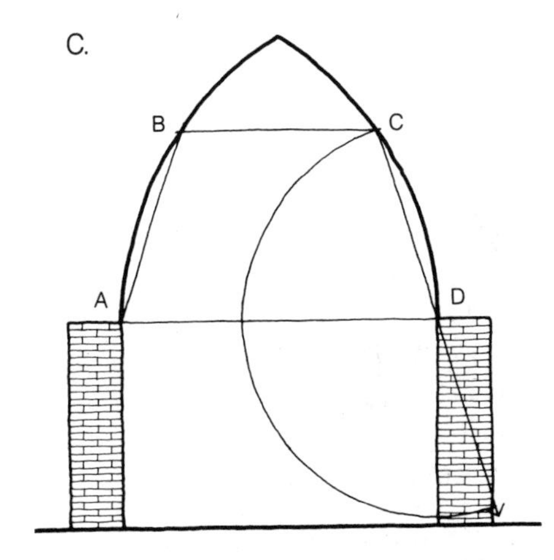

3.17 Lines of force in A. a barrel arch (half circle), B. a shallow arch (section of circle), and C. a parabolic arch.

also carry horizontal loads (the push that a vault or a dome creates). To carry this load, as discussed before, either the wall must be much thicker, or buttresses must be built behind the wall. Very thick walls are both costly and space consuming; thus buttresses are preferable, and also create a stronger structure.

Arched forms (arches, vaults, and domes) create horizontal forces that try to rupture the supporting walls. To counteract such forces, buttress walls or other structures are constructed. The shallower the arch, the greater the horizontal force, and the need for bigger buttressing. A tall arch, called a parabolic arch, transfers the horizontal load at its base to the ground.

The general proportion for a buttress wall is a triangular shape, with the long leg against the wall and the short leg on the ground. The number of buttresses used depends on factors such as the height of the walls, roof loads, and length of the building; and calculations can determine their size and location. In general practice a vaulted room has buttresses on the two side walls, and a domed room has buttresses on each corner. Chains, bars, and wooden poles connecting the arch supports at **spring lines** are traditionally used to take horizontal stresses in conjunction with buttresses. Smaller-span vaults could also be used at the end of the structure to work as the buttressing element. Finally, the more sophisticated earth structures almost always utilize the buttressing element for functional, symbolic, aesthetic, and other purposes.

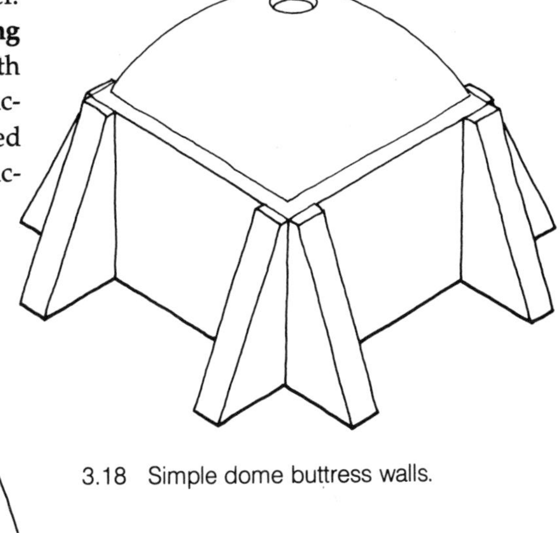

3.18 Simple dome buttress walls.

3.19 Simple vault buttress walls.

3.20 A flying buttress (known as *Posht band-e-Shamshiri*, "sword buttress," in Persian) in Yazd, Iran. Buttresses are structural, functional, and sculptural elements. A buttress can be in the form of a small vault or serve as a base for a staircase.

MUD-PILE (*CHINEH*) WALLS

The simplest way to build a wall is with earth, water, a shovel, and two people. Just wet the loosened ground, dig the mud, and pile it in rows about 40 centimeters (16 inches) high. Such a wall is commonly constructed in a tapered form when built around the property, and straight when used as a supporting wall. Mud-pile walls can be built without the use of any adobe block, and vaults and domes can be built on them.

These structures are certainly the cheapest and quickest to construct. Adobe is only used for the roof.

Mud-pile walls may be fired with the addition of bamboo, weeds, or other burnable straw-like material, which is laid across the wall to carry the fire through; or flues may be incorporated into the walls. The fire penetration is not as easy as in adobe walls. Like the rammed earth walls, mud-pile walls must be fired from both sides.

Tapered mud walls are built around the individual property, to protect a community or even to make long shadows for ice-making. Such mud walls over 8 meters (25 feet) long have been built this way and have lasted for centuries. These ice-walls are a great traditional technology of creating and keeping ice from winter to summer.

> Mentally, I design a very tall wall, like the ones they have been building in this country for hundreds of years and all of a sudden stopped building two decades ago—one of those yakhchal (ice pit) walls. Then, just as it was done before, I dig a huge, canyonlike pit at the south side of the wall and cover it with a roof a little above the ground. In the winter, in the shade of this wall and with abundant water, I make ice, lots of ice to fill up my giant-size ice pit.
>
> To this point, I almost follow the traditional ways step by step; but from here I depart in a new direction. I am not

3.21 *Chineh* (mud-pile wall) is the simplest wall construction. Mud is dug adjacent to where the wall will be; then it is loosened, wet, stamped on, and left overnight.

3.22 Two people construct the *chineh*. One tosses the mud up with a shovel; the other, resting on the dry row, piles it up in a fresh row.

3.23 Property walls are tapered *chineh* covered with mud-straw plaster, Iran.

3.24 Decorations are sculpted in mud or made with adobe blocks.

collecting ice as before, just to dig it up in the summer to make ice water, but I am building an entirely new type of cold storage house – a cold storage house that will contain thousands of tons of meat, fruit, vegetables, and perishable goods. A giant-size cold storage house that is built entirely around my giant-size yakhchal. A giant cold storage house that will need no automatic freezing equipment, no electric energy to run it, and will have no chance to fail any time of the year.

I start listing arguments against the modern and expensive cold storage houses built all over the world. I argue with the Western concept of energy–The problem with modern and Western thought about energy sources is that they are all on one track: fossil fuel, nuclear fuel, and the sun's rays are thought of as the only main sources of energy. Direct energy is the only energy people are concerned with. They worship the sun energy but forget the energy that is hidden in the shade of the sun. The shade, the cold, and ultimately the ice, are some of the greatest energies available to humanity.

3.25 Interior supporting *chineh* is built up straight inside, tapered outside for buttressing.

I then divide energy into two types: hot and cold. Hot energy we all know, but cold energy is what I have just discovered. Cold energy is energy multiplied by three: One, we use the fuel to make electric power. Two, we use energy to construct freezing and cooling equipment. Three, we use the electric power to run our equipment and make ice. In other words, we try ever so hard to change hot energy into cold energy. If we can make a shortcut and eliminate the three processes and still have our ice, then we make a great jump in energy savings.

Potential energy–snow and ice–is available and abundant in nature. Yet, like fools, we go our wasteful way of creating ice by all those three processes. Look at us. This unlimited, inexhaustible cold energy available to us in nature is completely ignored.

Look at us. Our cities all have power failures in the summer when we are in great need of cold air, and we spend fortunes making this cold energy, whereas we can just save it from the winter for the summer, for free. We can store it in winter, or even by bringing it from our snow-covered mountains to the heart of the city.

A piece of ice is a bundle of energy that can breed more forever–if we let the melted ice flow deeper into the layers of ice below. A piece of ice is more lasting than the sun; it exists even after the sun is extinct–eternal freeze.

Now I find a name for my new design–sardgah, "cold place," in Persian. Sardgah is the cold storage house built on top of a yakhchal, ice pit. It will be a huge place for hundreds of trucks to load and unload. It is a place entirely built on the energy given by nature to use, for free, for love, forever.

I start comparing the economy of a cold storage house that I build as a sardgah with the conventional one. The cost of construction? Less than one-tenth–cost of clay versus the cost of steel. The cost of operation, say for twenty years? Less than one per thousand–no electricity. And the chance of failure? None–no mechanical parts. One ice pit may last for a hundred years–by adding and breeding. And building the cold storage houses is just the beginning.

I walk, I think, and sometimes I squat and sketch some figures on the ground with my little stick. Yes, the sardgah for the cold storage houses is just the beginning of using cold energy. The sardgah system can be used to cool a building, or a complex of the buildings, and even an entire town in the summer, with this simple method:
Build a wall, use the shade in the winter, and make ice;

3.26 Arches, vaults, and domes can be constructed with adobe blocks over *chineh*.

store the ice in an underground ice pit; use wind catchers, small towers catching winds, to catch the wind and send it down through the ice pit to cool; then viaducts take the cool wind to the building; or just use the same pit for the storage on top of the ice.

The thought goes further, and I start sketching actual towns with these sardgah systems with many new uses of the sun, shade of the sun, and the wind. And then in a small sketch I use this tall, tapered wall for both the hot energy and the cold energy: the south side exposed to the sun has sun collectors that make hot water that is stored in the hot water tower. The north side makes shade and ice opposite the hot water tower, on the shady side is the wind tower to catch the wind and send it through the ice pit and from the underground ducts to the buildings. So it is easy to see how the new towns should be built now. The new criterion for the building of a town is this:

A new town for now and the future should be built first according to the sun and the seasons. The new towns should be built based on the availability of hot energy and cold energy—the availability of the sun, fuel, nuclear power, and all the hot energies from one side, and the availability of ice from the other side. The mild climate cities are not to be populated, but used as the resort places. Finally I come to the conclusions that Tehran, Paris, and New York are lucky, because they have cold seasons, and Bombay and Los Angeles are unlucky because they have only warm and mild seasons.

The globe cannot afford all the waste of energy that goes to the cities that have only natural hot energy, the sun, or natural cold energy, the ice, but not both.

C·H·A·P·T·E·R 14

HOW TO BUILD ARCHES, DOMES, AND VAULTS

The arch support and curved roof, of which domes and vaults are the main forms, is one of the greatest inventions of civilization. To learn how to build an arch, a dome, or a vault is to take advantage of the secrets and formulas our ancestors have left for us. I have visited the earliest recorded vaults still standing, in southern Iran, which date from 1300 B.C. They were thought to be the first ever built, yet new diggings in another area have turned up a vault roof dating back to 2000 B.C., perhaps earlier.

Obviously, curved roofs have several functional and structural advantages over flat roofs. This great invention was not only inspired by the "dome of heaven," as Omar Khayyam calls the universe, but from the basic need to cover a house with a roof in a land where there were no trees to cut and thus no wood available. The only materials at hand were the earth and pieces of rocks. To figure out how to build a roof over a shelter with these small units as the only construction materials, and one's hands as the main tools, took great human genius. Once done, humans went on to develop the sophisticated dome. (Some Persian poets suggest that domes were inspired by the shape of a woman's breast, and many of the dome roofscapes in Iran plainly give this feeling.)

A small dome, made centuries ago from small pieces of rock in a village in the south of Iran, was as great an engineering feat for its time as the geodesic dome of the American architect/philosopher Buckminister Fuller is for our time. Both work on the same principle of compression and uniformly distributed loads, and both are composed of small units.

Domes and vaults in general work with the hot and arid climates much more efficiently than flat roofs do, because they make sun and shade zones, catch the breeze, and create an inside air current.

Curved roofs also have some structural advantages. Their inherent

3.27 A domed structure built entirely of on-site rocks with lime-mud mortar, Fars province, Iran.

3.28 A geodesic dome, California Polytechnic State University, San Luis Obispo (a Geltaftan workshop is being held here).

strength is greater than that of flat roofs. Further, curved roofs can be built with small units and natural construction materials to gain the highest strength; while a flat roof needs large supporting members, which necessitates the cutting of trees. And, of course, arched roof construction lends itself to earth structures.

Once the technique of dome and vault construction is learned, roofs can be combined and used interchangeably in the most suitable and simple manner to create the strongest structures and the most beautiful spaces.

ARCHES

Architect Louis Kahn once said, "Ask bricks what they want to do and they'll say, 'Make us into an arch.' Try to sell them lintel and they'll say 'make us an arch.'"

The arch is a structure with a spirit; it is the structure that nature creates to be in tune with itself. An arch is a cut of a vault, a curve of a dome. And it is a simple and satisfying structure to build.

Arches created by nature, such as the Landscape Arch in Arches National Park, which spans 89 meters (291 feet); Pont d'Arc in southern France, 34 meters high (112 feet) and a 59 meter (194 feet) span; and La Arch, the great arch of Porta in northern Chile, all point to the natural limits of an arch structure in balance with the earth's gravity. Manmade arches—of the great mosques in the Middle East, the cathedrals in Europe, and the old Persian and Roman palaces and monuments—manifest the human quest to reach the limits of arch structure in tune with gravity.

An arch is a curved structure kept in balance by the pull of gravity. An arch is a curved structure that supports its own weight. Our mason, Ostad Asghar, showed us how to pile adobe blocks on top of each other, without form or mortar, to make an arch. To show the inherent strength of an arch, and also his skill, he held up some adobes in the air until they met at the arch point, then thickened the stems with a few more adobes, and the arch was made.

Adobe Forms

Here's a way to use the adobe itself as a form for an arch.

1. In a wall opening, such as a door or window, make a temporary stack of adobes, without mortar, as high as where the arch will begin. Then use smaller pieces of adobe and mortar as fillers to create an arch form.

2. Make the arch. The usual method is to lay one row of adobes flat over the form, with mortar in their joints, and then lay adobes vertically to create the arch. (The double-adobe width arch is formed this way. The upper flat lay can be eliminated.) The vertical adobes will have smaller

3.29 Arches two adobe blocks wide made with adobe block form work. Note the top of the wall stepping at the arch spring line to receive the first skewback adobe.

joints in the inner circle than the ones at the top of the arch. Use mortar to adjust differences in joint distance.

3. Remove the stacked adobe. If we fire the structure, we can take the adobes out after firing. Since they have no mortar, they won't fuse together.

Plaster Forms

You can pour a portable form right on the site using quick-set gypsum, plaster of paris, or any other quick-drying plaster or cement material. If our arch is small, as for a doorway, the form can be made in one piece; if the arch is large, it can be made in two pieces for easy handling.

1. Draw the top line of the arch right on the ground.
2. Draw a 15-centimeter- (6 inch-) smaller arch inside.
3. Lay two rows of dry adobes tightly next to each other on both the outer and inner edges of the arch. This will be our mold.
4. Pour some sand, straw, or sawdust, or lay a piece of newspaper in the bottom and edges of the form, so that the gypsum or plaster of paris won't stick to the ground or the sides.
5. Pour the gypsum or plaster of paris into the mold and smooth out the top.
6. To make two half-arch forms, for easy handling, put a piece of narrow board at the top.

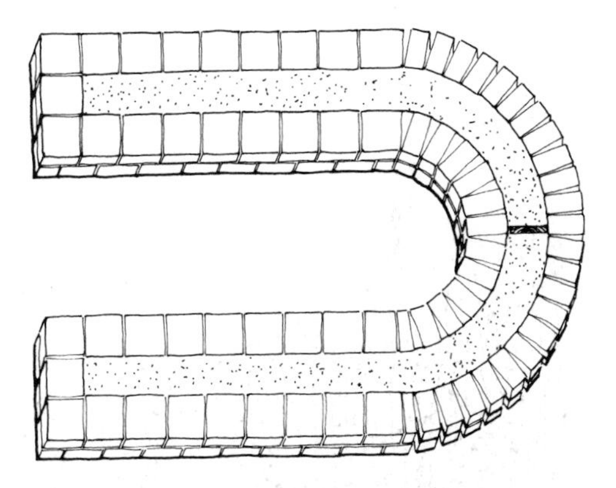

3.30 Casting arch forms. For an arch wider than a doorway, insert a dividing board at the top to cast two halves of an arch form.

After the plaster has dried (it will take about an hour), we can remove our arch forms. We now have a form about 15 × 15 centimeters (6 × 6 inches) thick, which will support our adobe arch while it is being built.

If we use a larger form, let us say over 1 meter (3 feet) wide, we can put some steel reinforcing bars, bamboo pieces, or long branches in the mold before casting our form for added strength. These unreinforced forms are usually used only once, since they may break.

For arches built with two rows of adobes, two forms must be made. Each one is secured under each row of adobe arch. Sometimes these molds are left in place, and become part of the structure. Of course, in this case the opening must be calculated accordingly, and better materials must be used.

Wooden and Metal Forms

Wooden and metal forms are usually used to save time and material and to create uniform arches. They can be used again and again for all openings of the same size, such as doors and windows. The usual material is plywood, but these forms can also be made of scrap metal. Using six wooden forms we built 144 semicircular arches – two bricks deep and two

3.31 Plywood or scrap metal can be used as a form to construct arches.

bricks high, with three meter (10 feet) span – in a large office building constructed with fired brick masonry near Isfahan, Iran. In the school project we built fifty adobe arches with only three sizes of scrap-metal forms.

To make the form:

1. Construct a truss to support the arched surface.

2. Bend a piece of plywood or sheet metal as an arched surface and nail or weld to the truss.

To construct the arch:

1. Wedge the form between the side walls of the opening, or stack two temporary (mortarless) columns of adobe, one on each side, under the form (or use wooden posts).

2. Construct the arch as described for adobe forms.

3. Pull out the form and reuse if needed.

3.32 An arch springing from a wall.

Arch Geometry and Construction

The curvature variations of arches, domes, and vaults have no limits. Historically used arched forms arose from human intuition, logic, culture, religion, or simply from copying nature. Some of the curves – semicircular arches, hemispherical domes, and barrel vaults – have been used more than others because of their inherent strength and simplicity of construction. Others have been dictated by the material and method of construction, such as earth block and corbelling techniques.

Here are some of the more common arch geometries:

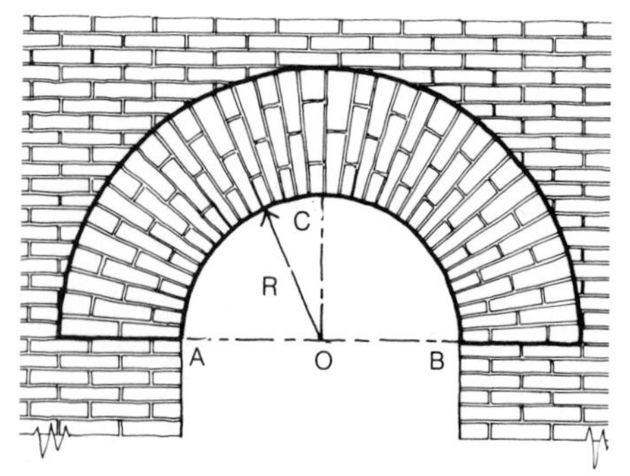

3.33 Semicircular arch.

1. *Semicircular Arch*. The arch span at the spring line is equal to AB, and the R radius is one-half of the span. With center at O we draw the half-circle ACB and build our arch, starting from the spring lines and finishing at the top. The mortar joint at the inner circle is minimum and increases to maximum at the outer edge of the arch. Adobe arches for door openings are commonly made two adobes deep (the thickness of the supporting walls) and one adobe high. For larger spans, two or three adobes deep and high is needed for strength. A two-adobe-high arch is shown in this illustration.

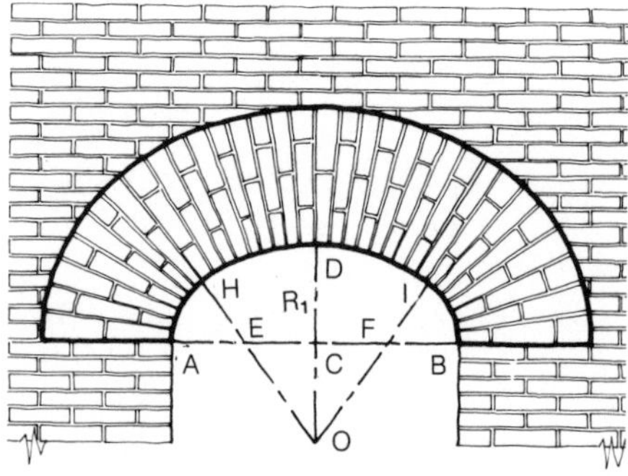

3.34 Four-section arch.

2. *Four-Section Arch*. The four-section arch (also known as the depressed three-centered arch in Europe), is a segmental arch with secondary curves at the spring lines. The center point O is arbitrarily chosen on the OD axis (the lower the O on the axis, the shallower the arch). The span at the spring line AB is divided into four equal sections, and OE and OF are connected and continued. An arc with OD radius will intersect with two small arcs generated by EA and FB radii.

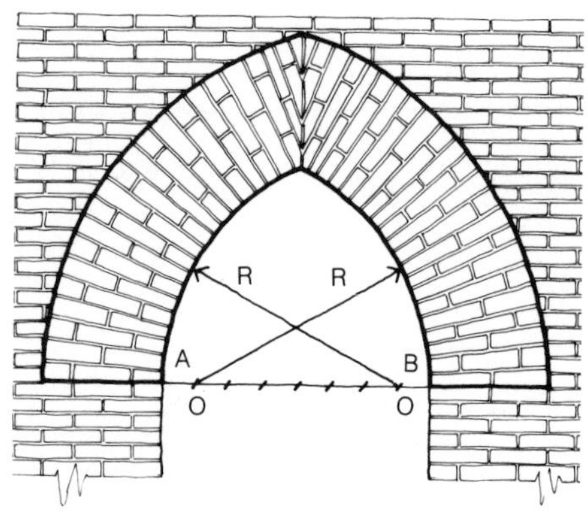

3.35 Eight-section arch.

3. *Eight-Section Arch*. The eight-section arch (known as the pointed arch) is similar to *shakhbozi* ("goat's horns" in Persian). The span at spring line AB is divided into eight equal sections. Two equal arcs of R from the O points, one-eight section, will intersect at the top of the arch.

4. *Segmental Arch*. The span at spring line AB is intersected with the axis line of OD at midpoint C. The rise of the arch CD is selected as desired, and the radius OD is determined by compass so that the arc intersects with three points of ADB. Since the arch starts with an angle at the spring line, the top row blocks at both sides of the arch are built in an angle to accommodate the first leaning blocks of the arch.

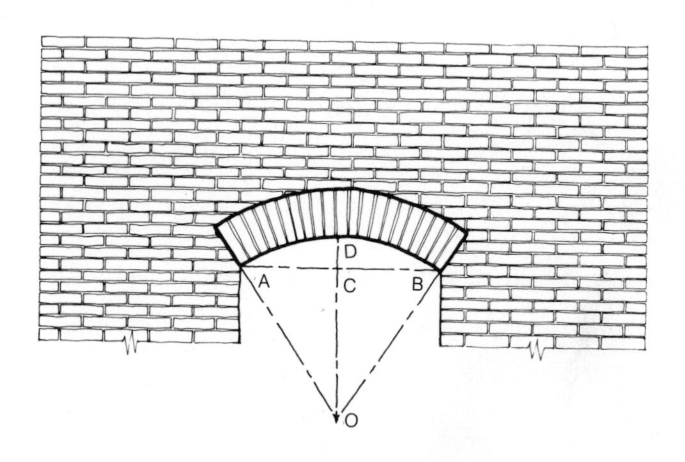

3.36 Segmental arch (the D–C rise can become almost zero, although some small rise is needed to prevent the optical illusion of flatness).

DOMES

We can build a dome without using wooden forms, centering, or any help from non-earth materials. It is a simple learning process with great spiritual rewards.

A dome is built over a square or circular room, or over a many-sided shape that approximates a square or circle (e.g., a hexagon or octagon).

The curvature and radius of the dome may vary, depending on the structure and aesthetics we want to achieve. High-profile, low-profile *corbelling* (corbelled domes are built of concentric rings of masonry blocks, and are tall and conical), simple, or double domes are built in relation to the total effect, or to achieve the largest spans. Generally speaking, a room size is decided based on the average span a mason can build with adobe blocks in common practice, and then a dome is constructed. Common room-size domes are usually made close to half-sphere geometry. However, a high dome or a very shallow dome can also be built easily, as long as the rules of balancing the loads with other domes or vaults, or buttressing principles are followed.

The pictures in this chapter show how one or two people can build a dome, continuing from underneath a dome until it closes. An area can be left open at the top for a skylight. The skylight can be built by making a curb around the opening at the top and fitting a ready-made skylight over it, or simply by making a curb with one side higher than the other and then putting a framed pane of glass over it; the skylight may be either fixed or openable. And of course an arched cupola, skylight, and air vent over the dome can be sculpted. To make a guide compass for the uniform curve of the dome and the angle of the inclination of the adobe rows, a simple pole and rope may be enough. Fix a pole in the middle of the room, say in a large sand-filled bucket. On top of the pole, and at the height of the wall, fix a rope. This will work as a compass. We hold the edge of the rope at every few adobes we have laid to make sure that we are not too far off. There are very detailed ways to make a compass that will lead at every course. An adjustable wooden compass with a lip to fit over the edge of the adobe is used in common practice, instead of a rope, especially for pendentive domes. With enough practice, we can use the reach of our arm as we would the rope, especially when building a small room.

To reach the top of the wall to build squinches or pendentives (interior corner supports), and domes, we can stand on empty oil barrels or boxes, or make a continuous scaffold in the room. We can then walk around parallel to the wall and inside the dome. The scaffolding can also be made with the help of the adobe wall itself. Fix four wooden poles into a notch made into the walls, and put four planks over them. The notch or hole in the wall can be patched and plastered later. Our mason built the entire 500-square-meter (5,000-square-foot) school using six empty oil barrels, four planks, four long poles, and two ladders for support.

A dome may be made in several ways, depending on the desired shape, and available materials and skills. The terms **squinch** and **pendentive** domes are commonly used, and often refer to the same tech-

3.37 All walls, arches, and *espars* (curved walls) are built and left to dry and settle for a couple of weeks before domes and vaults are constructed. To the left are the arcade arches and to the right are the rectangular rooms with *espars* at the end walls.

nique. To simplify the learning process, we will discuss both squinch domes and pendentive domes; **double dome** will also be briefly discussed.

SQUINCH DOMES

A squinch dome is a dome built over squinches. Squinches are interior corner supports – diagonally built, cupped-hand-shape arches, that hold up the dome.

Building a dome over a circular room is a simple and ancient technique. A dome is constructed by building concentric rings over the base by the corbelling method, which is to project each ring a little over the ring below – like steps in reverse. But building a dome over a square room needed human ingenuity. And that is when squinches were invented.

The words *sequnj* and *gushwar* are the old Persian words for squinches. According to some Iranologists, the word squinch is actually derived from *sequnj*; squinches are called *gushwar*, meaning earrings, because of their resemblance to women's earrings.

Squinch domes and pendentive domes are sometimes used in conjunction with each other; thus there is not always a definite use of one or the other, but many times a little of both. A squinch dome is usually built over a square or polygonal room, after the room's walls are finished. At each corner of the room, over the walls, a squinch is constructed to create the main supports for the dome.

To build a simple squinch:

1. Lay one adobe block, with mortar, diagonally over the top of the corner where two walls meet.

2. Spread mortar on top of this block and lay two adobe blocks over it. These two adobes touch each other at the top of the first block, and they also touch the top of the wall; thus a pitched arch is created.

3. Repeat the same work: Laying blocks over mortar in rows of arches that lean on each other and are slightly angled toward the inside at the top. (This angle will decide the height of our dome. It is created either by the help of a guide compass, a string, a template, or by eyeballing, and the extension of our arms.)

All adobe blocks touching the top of the walls should be in line with the interior edges of the walls. The point to remember is this: All adobe blocks touch each other dry—without mortar—at their vertical joints. Mortar is only used in laying adobe blocks over each other, but not next to each other. When building leaning adobe arches for domes or vaults, the empty space between the two adjoining adobes is filled with dry pieces of broken bricks or rocks, which is called a wedge or **dry-packing**. Every joint does not need a wedge.

To build a squinch dome over a square room:*

1. Build walls on the four sides of the rooms to the desired height, with the door- and window-opening arches in them. Opening arches may be flush with the top of the walls or rise above them. Leave the constructed walls alone for a couple of weeks to dry and settle.

**For more details see Part 5, Model-making.*

3.38 Squinches are built at four corners. To begin a squinch, an adobe block is laid over the wall at the corner with mortar at a forty-five-degree angle. The block projects out a little into the room and tilts inward. Two adobe blocks are pitched over the first one with mortar. The bottom part of the blocks must touch the top of the wall. Squinch domes and vault construction stages are from the Javadabad Elementary School, Varamin, Iran.

3.39 When the process is repeated, each layer being tilted a bit toward the inside, the squinch curvature is naturally created.

2. Build four squinches at the four corners, large enough so that the base of the last adobe rows of adjacent squinches meet each other at the halfway distance of the walls.

3. Continue with the last rows of blocks making the squinches, until the dome is finished at the top; the top rows will create a square opening (continuous squinches).

An alternative route to step 3 is to construct rows of blocks to fill the

3.40 When squinches meet at the center of the wall, continuous rows of blocks are laid over them to close the dome with a square opening at the top. Note the dome base: one corner is a squinch and the other a semi-pendentive. (An alternative method is to build up a circular base, and then close up the dome with concentric rings, as shown in figure 3.43.)

3.41 Finished arcade, Javadabad Elementary School.

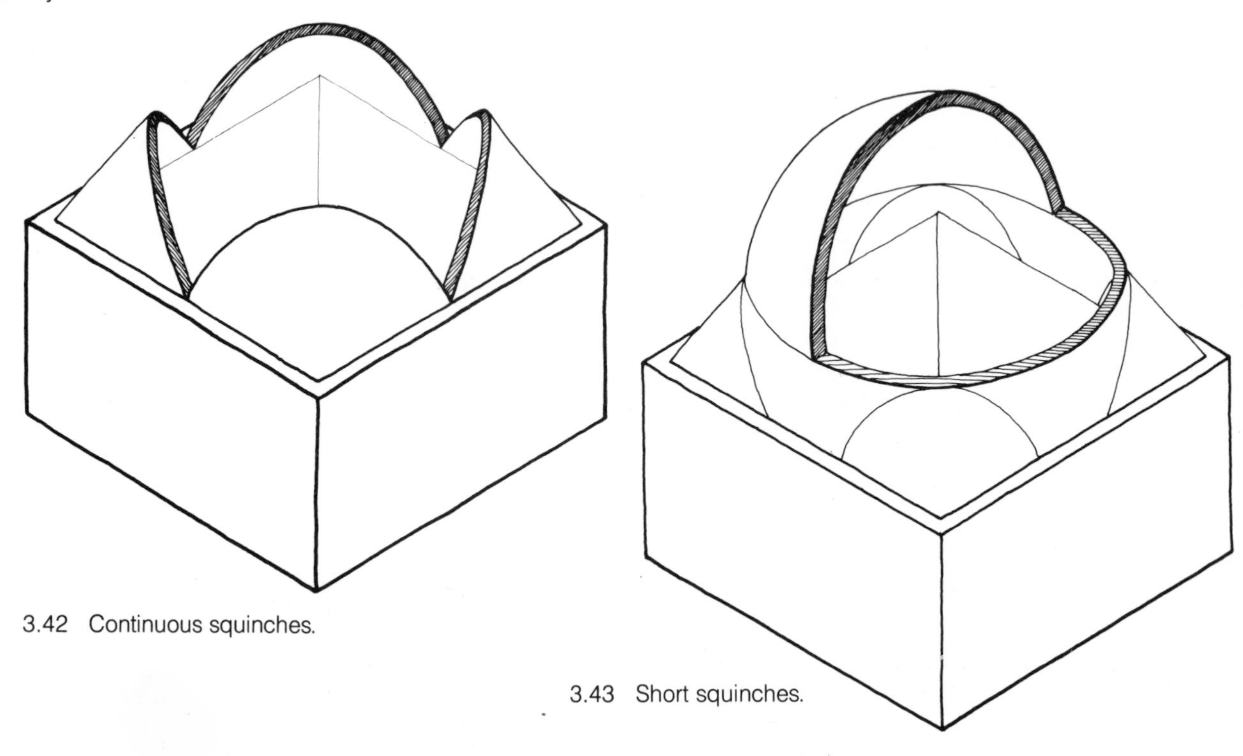

3.42 Continuous squinches.

3.43 Short squinches.

gaps between the adjacent squinches until it creates a circular base at the top. Then lay rows of blocks in concentric circles to close up the dome (short squinches).

Once we have learned to build a dome on a square base, exactly the same steps can be carried out to build over a hexagon, octagon, or similar shapes. For an octagon, for example, we would build eight squinches instead of the four we would build on a square. The eight corners rise and become a dome.

There are many different ways to construct a dome as far as the patterns of the blocks are concerned, but basically they all try to achieve the same result: a roof structure built with small material modules to transfer all loads to the ground in compression. By building even a small dome or a vault, we will give birth to a new consciousness, and the round structures of the past or the future will have new meanings for us.

PENDENTIVE DOMES

The pendentive dome is really a refined squinch dome. As the walls supporting the squinches and the dome are replaced by arches, the term pendentive is used.

A pendentive is a triangular piece of support that springs from the corner of the room to hold up the dome. It also transfers the weight of the dome directly to the four corners of the room.

To build a simple pendentive dome over a square room, we must first build four adobe arches, one on each side; we then fill up the open shoulders of the arches with four pendentives until the pendentives meet to create a circular (or polygonal) base. The dome is built over this circular base. We lay rows of adobes in concentric circles that gradually move inward. The sequence for building a pendentive dome over a square plan is as follows:

1. Build the foundation walls to the finish floor level.

2. Build four arches on four sides of the room to meet at the corners. The arches start on top of the foundation walls.

3. Build four pendentives to close the shoulder-gap of the connecting arches.

4. Continue the last rows of the blocks, making the pendentives, until they meet and make a circular base.

5. Lay rows of adobe blocks in concentric circles, stepping in a little at a time and tilting, following the guide compass, till the dome closes at the top.

When domes and vaults are combined, their slopes and forces can be used as the supporting elements for each other. Even though pendentive domes are used frequently in many parts of the world, the squinch domes, in small-scale work, may be constructed just as easily.

3.44 Different stages of building pendentive domes are shown at the Dar-Al-Islam complex, Abiquiu, New Mexico. Guide compass, made with wood, is secured in a barrel and located at the center of the room. The mason walks over a continuous scaffold and lays the concentric circles of the dome.

DOUBLE DOMES

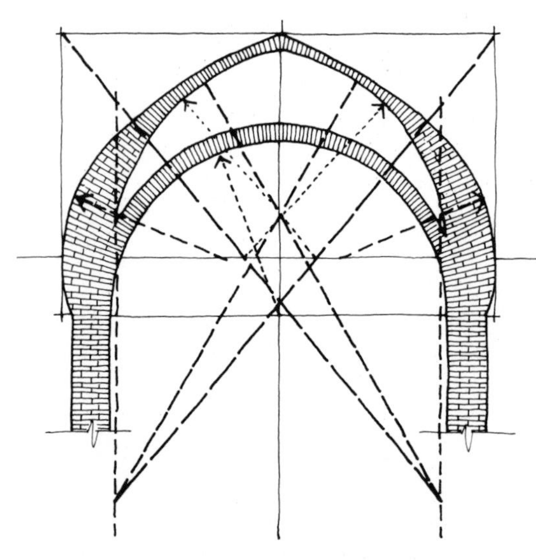

3.45 A double dome with outer dome projecting beyond the supporting walls. The outer dome is integrated with the inner dome at the base to counterbalance the stresses.

Double domes are sophisticated domes that are structurally stronger, functionally more suitable for climatic control in extreme weather, and artistically more flexible in design possibilities. The outer dome takes the sun and the cold, the sandwiched air space works as insulation, and the size of the inner dome does not fluctuate with temperature. The inner dome is built to the desired ceiling height and can act as a structural shell independent from the outer dome. The outer dome in many cases sits on the base of the inner dome and works as a buttress to counteract the tension. Both domes may be connected with radial fin walls, built in the air space; or they may be free from each other except at the base and the top.

The shape of the outer dome is based on aesthetic rather than functional considerations. Because the outer dome is not in contact with the interior space, there is less expansion and contraction; thus longer-lasting roofing finishes are possible. Double domes have been used in public buildings and larger spans and are mostly constructed with fired bricks. The thickness of the dome shells decreases from the base to the top. There are many large-span double domes constructed with masonry in Iran, which have lasted for centuries. Double domes built without form work; to achieve the ultimate in structural harmony with gravity was the master builder's quest. Using squinches, a square room was first changed to an octagon, then to one with sixteen sides and finally to a circular base for a large dome.

Because double domes require more skill and material to build, they may not be fit for the owner-built low-cost housing we are discussing. However, they are very energy efficient and can work wonders in extreme climates. If shortcuts in building small double domes can be found, the contribution to earth architecture in harsh climates will be of great value.

VAULTS

Vaults are even simpler to construct than domes. A dome is a double-curvature form (it curves up and down and it curves sideways), but a vault is only a single-curvature form (it curves up and down). A vault may be thought of as an arch built very deep; a dome may be thought of as an arch rotated around itself. The great secret in building a vault is to construct leaning arches—build an inclined arch first, and then construct more arches leaning over the previous one.

Even though a catenary or parabolic arch creates the strongest vault structure, such curves are not always desirable because of the very high ceilings they form. For simple spans such as common-size rooms, much shallower vault roofs can safely be built, as long as they are buttressed.

A catenary curve is an arch higher than half a circle. The curve for this arch may be found by hanging a chain between two nails on a wall. This technique is common practice among potters who make their own kilns. A parabolic arch is close to a catenary curve. It may be drawn with the following steps:

1. Divide one half of the arch-base span to equal segments.
2. Divide the total height of the arch into same number of segments.
3. Draw parallel vertical lines from the base segments.
4. Draw parallel horizontal lines from the height segments.
5. Draw the arch curve by connecting the intersecting points.

Vaults have two main supports—the two legs. A dome is supported on four or more sides, but the main support for a vault is the two side walls. However, to start building a vault we must begin from a third wall, an end support.

Vaults are built over long or rectangular rooms. The end wall of the room is where the vault begins. To build a vault, the following steps are usually taken:

3.46 A curved wall (*espar*) with window opening at the center.

3.47 Layers of adobe blocks are laid flat, parallel to the window arch, to create the *espar* curvature, which is the same as the vault curvature. Here rows of *espars* are ready for vaults.

3.48 To create a skylight, a vault is constructed from each end of a room and the bottom of the gap is filled with horizontal rows of blocks.

3.49 All vaults are begun and built in sequence to counterbalance each other's forces. (This will eliminate the need for buttressing during the construction.) Note that all arches making a vault are leaning arches; Javadabad Elementary School, Varamin, Iran.

3.50 A mason and a helper can build a vault standing on a plank. The helper usually rubs the mortar on the layers. Someone else tosses up the adobe blocks from below. Here a son is handing his father the wedges while learning his father's skill.

1. Build four walls of a rectangular room to the desired height, including doors and windows arched openings.

2. Build an ***espar*** (a curved end wall) to the desired height of the vault.

3. Build a vaulted roof, starting from the *espar* wall and moving forward in rows of inclined leaning arches.

Vaults are often built in rows because they are stronger and more economical. Thus the way to build them is not one vault at a time, but in sequences of around 1 meter (4 feet) deep to counteract each other on top of the walls. The only tool we need is a plank under our feet. Planks can be laid over the walls, over poles stuck into the walls, or over the rungs of two ladders leaning on the two side walls. Sometimes an adz is used to break the adobe; but with practice we, like the skilled mason, can simply break the adobe with the heel of our hand or with another adobe.

And instead of tapping the adobe in place with an adz, we can simply strike the adobe with force to set it into the mortar (this may make the mortar splash).

Now let us review the process in more detail.

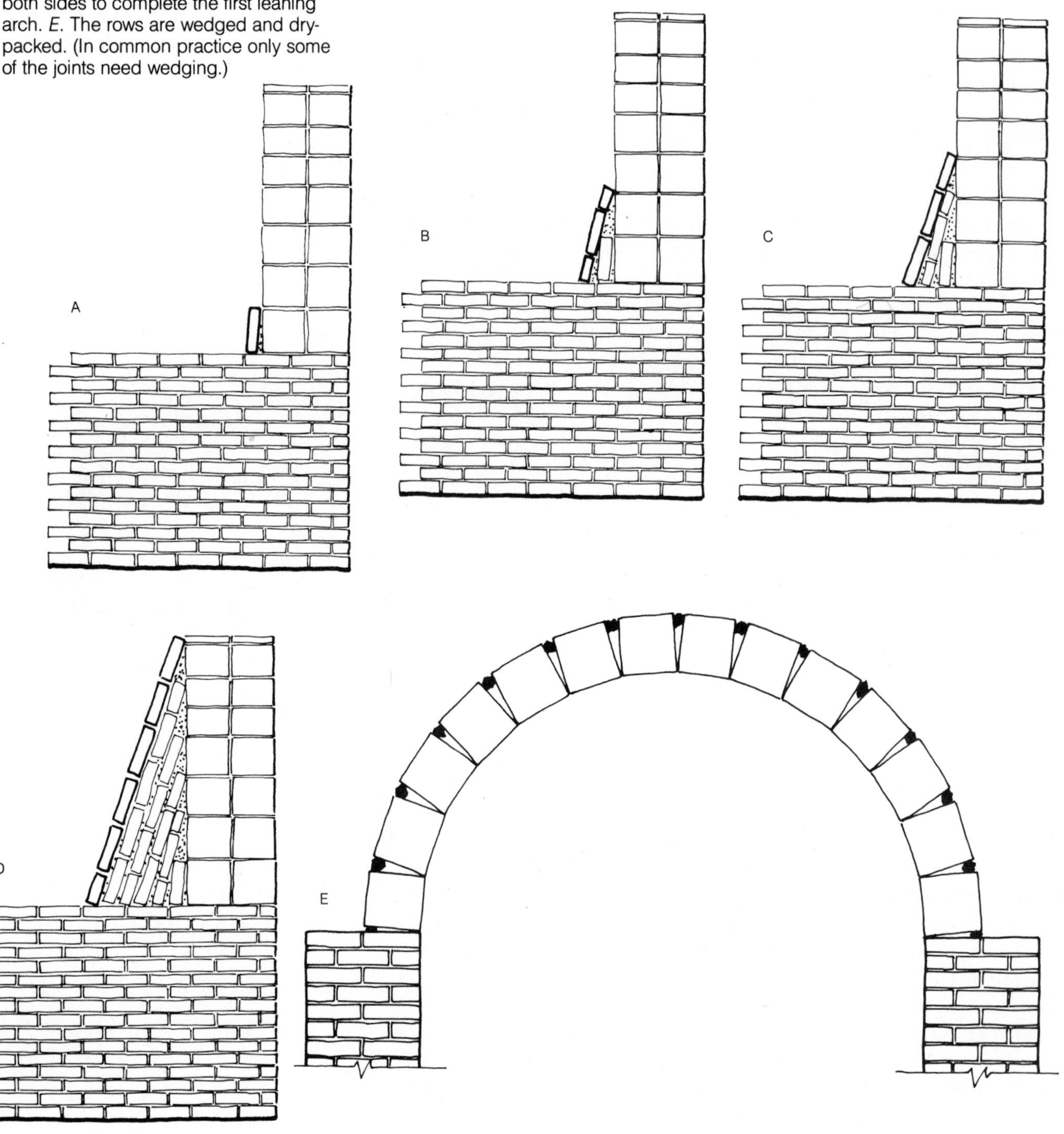

3.51 Building a vault. *A.* One adobe block is laid vertically against the *espar* over mortar. *B.* The second row starts with a half block; a full block is pitched over the previous block to touch the *espar*; another half block tops it. *C.* The third row starts with one full adobe and continues over the inclined angle. *D.* Rows continue from both sides to complete the first leaning arch. *E.* The rows are wedged and dry-packed. (In common practice only some of the joints need wedging.)

1. Build all the walls in standard fashion. If possible, avoid cutting door or window openings in the side walls, which carry the vault load. A wall 2 meters (6½ feet) high will provide ample room space once the curved ceiling is added. The tops of the arched door and window openings can be flush with the top of the walls; or the arch may project above the wall depending on the wall height. (This is true for both dome and vault wall bases.)

2. Build the *espar. Espar* is the Persian term for a curved end wall of a vault. The first rows of vault-arches will lean on it.

The easiest *espar* to build is one without an opening. To do this, simply build the end wall so that it ends in a curve similar to the vault curve.

To build a curved wall over a regular height, two-adobe-thick wall, lay two adobes on top and in the middle of the wall. Lay the next adobes pitching over the first ones. As the successive layers are constructed, the small pitch will automatically change to curves, thus making the curved end wall.

Another way to make an *espar* is to build an arch at the door or window opening and then lay the adobe blocks over it to make the end curved wall. The distance between the adjoining *espars* on top is used for roof drainage. *Espars* are always made at the same time as the rest of the end wall to ensure a good bond.

Yet another way to make an *espar* is to build the end walls straight up, ending a little higher than the vault curve. Draw the curve of the vault on the wall with a piece of chalk, and apply the first layer of the mortar following the inner curves of this line. This is easier to do if the design specifies a taller end wall as part of the roof spandrel. In this design the straight back wall hides curved forms of the vaults.

Even though the vault is built up from and leans on an *espar* or the end wall, the load of the vault rests only on the two side walls and the *espar* carries no load.

3. Begin building the vault. Lay one adobe vertically against the *espar,* over the mortar. The second row starts with one half-adobe and, with a full adobe, is pitched over the previous block to touch the *espar.* The third row starts with one full adobe and goes over the inclined angle. Continue in this manner with the next rows, until the first arch is completed between the two side walls, leaning against the *espar* or the straight wall. The first arch is usually completed in the first five to seven adobe rows, depending on the angle of the inclination and the span. The inner row of the *espar* wall can be curved to become the vault itself, depending on the skill of the mason.

4. As the adobes are laid up to the top, their ends touch each other. Adobe blocks on the inside of each row of arches are set tightly next to one another; but in the outer circle joints are open between adobes, and may need wedging (dry-packing).

5. Wedge (dry-pack) the blocks. This is, put pieces of rock or fired brick in the joints so that the adobes don't slip down. (In fired structures, these wedges must be either hard adobe or fired bricks, since rock pieces will break apart in the fire.)

Dry-packing is very important—without it, the vault will not stay up in the air easily. A Persian saying shows how important the wedging is: *Bandeh ra Khoda negah midarad, tagh ra band(eh).* That is: The human is held up (in this universe) by the God, and the vault is held up (in the air) by the wedge. In common practice only the joints near the top of the arch get the wedges, others are filled with mortar.

6. A vault is built from both sides, and the last adobe is put at the top like the keystone of an arch. The length of our arm can be the compass, and the height of the vault is usually determined by the height of the scaffolding under our feet plus our own height. Once we know how high we must put the last adobe at the top—this will be the highest comfortable reach of our arm—we just follow the routine. Every row uses the same number of adobe blocks, with staggered joints. To adjust and continue the uniform vault straight, strings could be pulled at the top. But some practice is needed. The best thing to do is to practice with smaller

3.52 A combination of squinch domes and vaults.

3.53 A combination of a pendentive dome and a short vault (the vault is formed by multiple arches in this case).

models before building the actual vault. Sometimes wooden guide forms are used for speed and uniformity, but what is needed is some practice, not forms.

7. To close the vault at the end, continue it to the other end wall or *espar,* and then close it in by filling the gap with adobe and mortar.

8. To make a skylight, we build the vault until we reach the place we want the skylight to be. At that point, close in the two sides with adobes until the desired opening is reached. At the top of the roof, we make a curb with adobes to receive the skylight.

9. The end of the vault can also be curved in toward the other *espar,* arch, or however the end is, to make a semi-curve interior finish. Or the end of the vault could become a semi-dome when it reaches the end wall.

Some classical Iranian arches, vaults, and dome combinations, Sassanid Period (ca. A.D. 224–640).

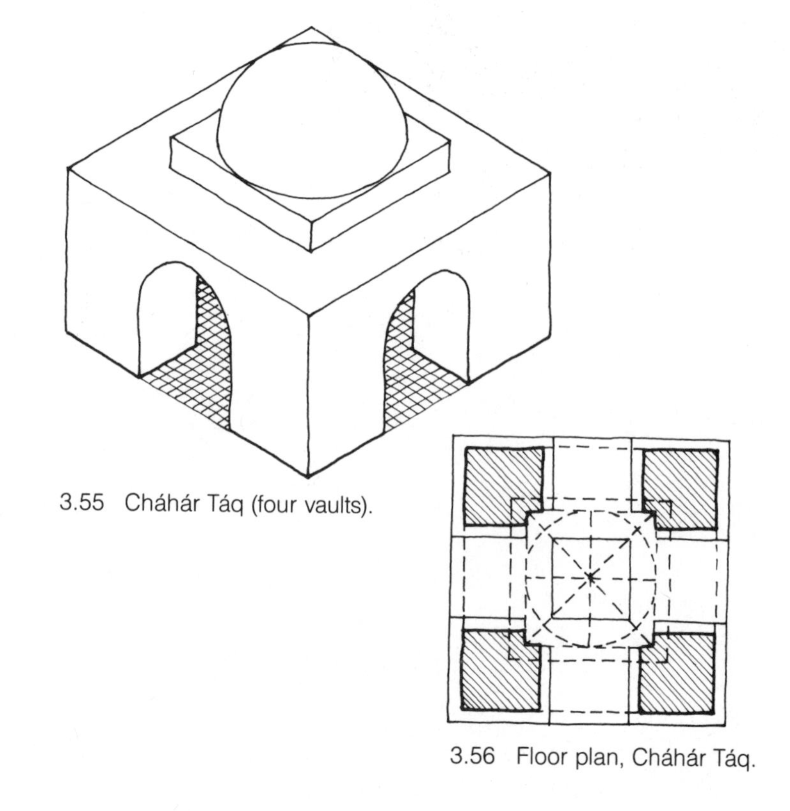

3.55 Cháhár Táq (four vaults).

3.56 Floor plan, Cháhár Táq.

3.54 Sarvestan Place.

3.57 Ivan-i-Karkheh. Massive arches are first constructed and then transverse vaults are built between the arches. (High windows are at two ends of the vaults.)

C·H·A·P·T·E·R 15

ROOFING AND WALL COVERING

Once the walls and curved roofs have been built, we must think about the roofing and exterior wall covering.

GRADING

Before we can waterproof the roof, we must grade it. The plan for grading and waterproofing should follow the general philosophy for all adobe buildings: Keep water away from the building. That means to grade the roof so that rainwater or snowmelt runs away from the building rather than toward it. Try to avoid gutters and downspouts, which drain between inner walls or columns that may seep through. Drain the roof water to the outer spaces or to the open courtyard and lead it away from the building. The rainwater or groundwater around the building must also be drained through gravel trenches, perforated pipes, storm drains, or the like.

Grade the roof by packing clay-earth between the domes and vaults to get the proper slope for rainwater drainage. We can dump the clay-earth material into place and roll a small hand roller over it; or we may pack it with a heavy wood rammer; or we may pack it simply by walking over it thoroughly.

WATERPROOFING

After the grading is done, the roof is ready to be waterproofed. There are many ways to apply roofing or waterproofing over the roof surfaces and exterior walls.

Mud-plaster

The simplest system is to use the adobe mix itself. A high-percentage clay-earth mixed with straw is one of the cheapest and simplest ways to cover the roof and the walls. The clay-earth absorbs water very slowly; and once it is wet, it does not let any more water pass through. When we see an area of land with puddles of water after it rains, this tells us that there is lots of clay in that earth. As a matter of fact, some of the best waterproofing materials made today, such as manufactured panels or sprayed-on waterproofing, are often nothing but clay. So use of the clay that is present in our adobe can be the base of the waterproofing system.

Mud-straw plastering often involves community participation. In Native American villages made with adobe and plastered with mud-straw, such as Taos Pueblo in New Mexico, the replastering is traditionally done by expert and skillful women. In the desert villages of Iran men and young children work as a crew, stamping and walking on the mud-straw until it mellows. They let it set for a week until a "juice" appears, and then they apply it. Mud-straw plaster has a dimension manufactured materials lack. The material and process has roots in the culture and the language. In fact, pregnant women of Iranian desert villages often crave the smell of the mud-straw plaster, especially in the rain.

In countries where materials such as asphalt, tarpaper, and cement are readily available, more permanent waterproofing can be applied. Using a permanent plastering and waterproofing material has its own validity, it is economical, time-saving, and maintenance-free. But mud-straw plastering is a community project. In societies that have plenty of time, being involved with each other and helping each other is a way of life, and a house is treated as a living entity.

To make mud-straw plaster, mix the clay-earth with straw in a ratio of 3 parts earth-clay to 1 part straw in volume, and let it sit for a few days to a week. This will create a good plaster. For a fine finish, mix clay-earth

3.58 The first coat of mud-straw plaster being applied over a dome with a trowel.

3.59 Roofscape of domes, vaults, and skylights at the Dar-Al-Islam mosque complex, Abiquiu, New Mexico, designed by Hassan Fathy, Egyptian architect.

and straw-dust (very finely ground straw). This mix, which is mostly used for interior plastering, gives a buff adobe finish. It may either be used for the whole wall or only for the trim and borders.

In the areas with freezing winters, a small amount of salt should be added to the roofing plaster mixture. The salt will lower the freezing point in the mud-straw plaster and will prevent it from freezing and cracking. The straw in the mixture works as a reinforcement and keeps the plaster from shrinking and cracking. It is also water repellent and creates better insulation against the heat and cold.

Mud-straw plaster for roofing is usually applied in two coats. The base coat fills all the cracks and joints and smooths out irregularities. After the first coat is dried out, the second coat is applied as the finish work. The second coat should have a greater percentage of straw. It may have to be reapplied every few years, depending on the amount of rain. The better the plaster is made, the longer it will last. A low-clay, low-straw plaster must be redone every year, while a high-clay, high-straw plaster, well mixed and mellowed for several days before being applied, may last several years. A stone roller is often used after the moisture from a heavy rain or snow has dried out, to compact the plaster on the roof.

3.60 Roofscape, Yazd, Iran. The roofs are covered with fired brick and mud-straw plaster.

Asphalt or Fired-Brick Roofing

Asphalt roofing or tarpaper and fired-brick roof covering is applied to curved roofs in much the same way as on sloped or flat roofs, depending on the region and the amount of snow and rain. The common practice is to apply a first coat of plaster and then apply hot tar and burlap and a layer of asphalt. Square fired bricks laid over the tar and burlap with a good mortar can create a handsome and permanent roofing.

Exterior Finishes

We can apply one coat of mud-straw plaster as the base and a coat of gypsum plaster as the finish. We can also use two or three coats of cement plaster on the wall for waterproofing and finish. For a more durable result, the cement plaster must be applied over a wire mesh. Even with wire mesh, however, the plaster and the mesh may separate from the adobe or crack during severe temperature changes, because of the difference in expansion rates of the cement and adobe. Systems that use gunnite or some other texture instead of mesh may also be used, if the material is available and proper application is assured.

3.61 Building with courtyard, Yazd, Iran. The roofs and walls are covered with mud-straw plaster.

3.62 The walls of the Dar-Al-Islam mosque, Abiquiu, New Mexico are covered with cement plaster over wire mesh.

A fired structure can be roofed like a common adobe structure. It may also be covered with fired brick or ceramic tile.

Mud-straw Plaster

If the mud-straw plaster is applied over the roof before the structure is fired, it will crack when the steam tries to escape during firing. The

3.63 Jamī Mosque, Yazd, Iran. The entry porch, minarets, and dome are covered with ceramic tile.

appropriate method is to try to bake the plaster with the structure at the end of the firing period. The following steps must be taken:

1. The mud-straw plaster is made from the same material used for adobe or the common mortar–clay-earth with very few impurities–and should contain a small percentage of straw.

2. Apply the plaster with a trowel in a layer over 1 centimeter (½ inch) thick during the last two to three hours of firing, when the steam is completely out of the structure and the roof is dry and hot. Even though the roof is hot we can stand on the walls, in the valley between the roofs, and apply the plaster over the dome or vault. This way the plaster dries and can be baked with the roof. The baked plaster will then cool off at the same rate as the structure to create a strong bond.

3. After a few days of cooling, apply the second layer of the mud-straw plaster. This layer is the same as the common plaster, and must be reapplied every few years, depending on the amount of rain or snow in the region. (Instead of the second layer of mud-straw, a more permanent roofing, such as tar or asphalt, can be applied over the first baked layer.)

To keep the insulating quality of the adobe, which is better than solid fired brick, we may want to create a composite structure–a fired structural shell within and an adobe shell outside. In this case the roofing treatment can be done as previously described.

Fired Brick or Ceramic Tile Roofing and Wall Finish

The roof, whether or not it is fired along with the mud-straw plaster, can be covered with fired bricks or tiles. These products can be baked inside the rooms while they are being fired.

When the structure has been fired, apply a layer of tar waterproofing over the roof. Then, with a clay-lime-sand or cement mortar, set the fired brick or tile.

Fired bricks or tiles can also be used to cover the outside walls. For the firing of the exterior walls, a secondary shell must be created to cover the outside wall and access must be provided to lead the fire from within the structure. To fire the roof structure all the way through, put down a layer of adobe blocks, flat and without mortar, on top of the roof. After all the steam has left the structure, apply a layer of mud-straw plaster over the joints of the flat adobes. This creates a kiln-like layer that allows the fire to penetrate and bake all the way to the outside of the roof structure. (We may use a fire-blanket or fire-insulated panels instead of the temporary adobe layers. Such materials are frequently used for ceramic kiln linings and are available in industrial societies.)

C·H·A·P·T·E·R 16

FLOORING, INTERIOR FINISHES, OPENINGS AND UTILITIES

FLOORING

Common Adobe Flooring

In a common adobe building, a lime-clay-sand mix slab or conventional fired brick or tile blocks may be used for flooring. Concrete slab floor, when used, is generally poured at the same time as the foundation. Plain clay-earth flooring, sometimes called adobe flooring, can also make an attractive and durable flooring if it is well maintained with sealer.

To construct an adobe floor, lay a pattern of wooden strips or large square panels on the floor and with them the clay-earth mixture. After drying, repair the cracks and coat the floor with two or three applications of sealer oil. (In some areas, such as the southwestern United States, animal blood and oil have been used both in the mixtures and the sealant.)

Geltaftan Flooring

In a fired structure the floor can be constructed and fired with the structure. The adobe layer is set on the packed clay-earth, and the flue openings are located below the slab level, so that the floor is baked before the fire is vented from the flues. To let the fire through the floor at the points of the flues, we can either leave an adobe-brick-size hole in the floor, to be covered with the same flooring later, or provide a duct space around the room under the floor. This duct space may be used later for plumbing and electrical wiring. If the structure is designed with a wind-cooling system, the duct space and vent shafts can be used later to circulate air through the structure.

The flooring can also be done after the structure is fired. This system may be more advantageous when the floor area of the room is used to

3.64 Javadabad School, Varamin, Iran. The east and central arcade interiors were covered with gypsum plaster after firing.

bake adobes, tiles, and pottery at the same time the building is being fired. After the products are taken out, the floor can be covered with fired adobe or tile or even wood flooring. The unfinished floor may also be utilized for plumbing and utility lines. The sub-flooring can be the natural ground with a layer of rammed clay-earth over it, which will harden during the firing. When the finished flooring is done afterwards, the flue openings can be at the natural grade level.

3.65 Interior Mirror Palace, Yazd, Iran. The walls are covered with gypsum plaster and the plaster carvings are with *gach-e-koshteh* ("killed plaster").

A general rule in firing the floor is to locate the flue openings in relation to the floor elevation. The lower the flue openings, the deeper down the fire penetrates.

INTERIOR FINISHES

Common Adobe Finishes

The interior finish of adobe walls and ceiling may be left exposed or covered with adobe material, gypsum, or cement plaster. The interior plaster is usually applied in two coats: Mud-straw plaster is the base coat, and gypsum plaster is applied for a finish. Both coats may be gypsum plaster; but the mud-straw plaster base coat gives a less expensive and more homogeneous finish.

To use gypsum plaster for ceiling and wall decorations, such as carvings, we must keep it wet and alive. Paradoxically, to keep the plaster alive we must first kill it (*gach-e-koshteh*, in Persian). To kill the plaster we must make it into a doughy paste and knead it for over half an hour. Such a mix will stay pliable for two to three days.

Adobe plaster as an interior finish (generally in a buff color) is made with a mixture of straw-dust, fine sand, and clay as the finishing coat. This finishing material can be used for the main plaster surfaces, while the white gypsum is used for the trim and borders. Such natural and attractive adobe finishes have lasted for centuries.

In plastering with cement (stucco), with or without wire mesh, two coats instead of three may be used. Galvanized wire mesh is nailed to the adobe on these types of finishes, and local building codes specify the materials and their applications. For interior plastering, the use of wire mesh is not necessary.

The easiest way to finish the interiors is to pay careful attention to construction details, and then whitewash directly over the adobes.

Geltaftan Finishes

Fired adobe and clay structures can become a dream work for graphic artists, interior designers, and sculptors. The interior spaces can be

3.66 Dar-Al-Islam mosque, Abiquiu, New Mexico. The interiors are covered with plaster.

sculpted to any desirable form before firing and glazing. The fireplace, bookshelves, and even the coathooks, light fixtures, and counters can be sculpted and glazed. (The system of firing and glazing is discussed in Part 4.)

DOORS AND WINDOWS

The rough frames for doors and windows are usually put in during wall construction. When arched at the top, the frame can be used as the form for arch construction. Later it will receive the door or the window. To secure frames to the walls, wooden adobe-sized blocks (called "**gringo blocks**") can be laid up with the wall, at two or three points of the opening height, for the frame attachment. The gringo blocks will burn out if the structure is fired.

A rough frame of either wood or steel is commonly used in many parts of the world. Steel frames must have an application of rustproof paint before installation. Finish plastering is joined with the rough frame, and the door or window is installed.

In fired structures, the rough frames are installed after the firing. Notches for the frame anchors can be made in the wall before firing; after firing, the frame is set with the anchors into the notches, then leveled and plastered.

Since the adobe walls are thick, double doors and windows can generally fit in the thickness of the wall without projecting into the room. Standard-size doors (90 centimeters, or 3 feet) if cut into two pieces (45 centimeters, or 1½ feet), can create an intimate human effect within the adobe dimension of arches and curved ceilings. And of course, in areas of the world where wood is scarce, smaller panels may be made with leftover wood scraps. There is no law against using square-top doors and windows; however, curved door and window transoms or curved door tops create a harmonious look with the arches and curved ceilings.

PLUMBING AND ELECTRICITY

The existence of piped water and electricity is a luxury most people of the world can only hope for. A common house built in most parts of the world has hardly any plumbing except for a water line to the toilet and a general use area faucet. Electricity consists of a general lighting lamp and a plug outlet for each room. The same goes for the sewage system. In the West and other wealthy societies, however, these utilities are the basic necessities of daily life and are almost always available. So designing and constructing plumbing and electrical lines can range from providing a single pipe or wire to the most elaborate means of comfort. The basic and vital need for a hygienic living environment usually can be met with simplicity, which the poorest people may easily afford. The location of an outhouse with proper regard for wind direction and underground water, and the provision for a simple gooseneck trap in a toilet

3.67 Sultan Mirahmad Mosque, Yazd, Iran. The interior details are of brick and plaster with a mud-straw base.

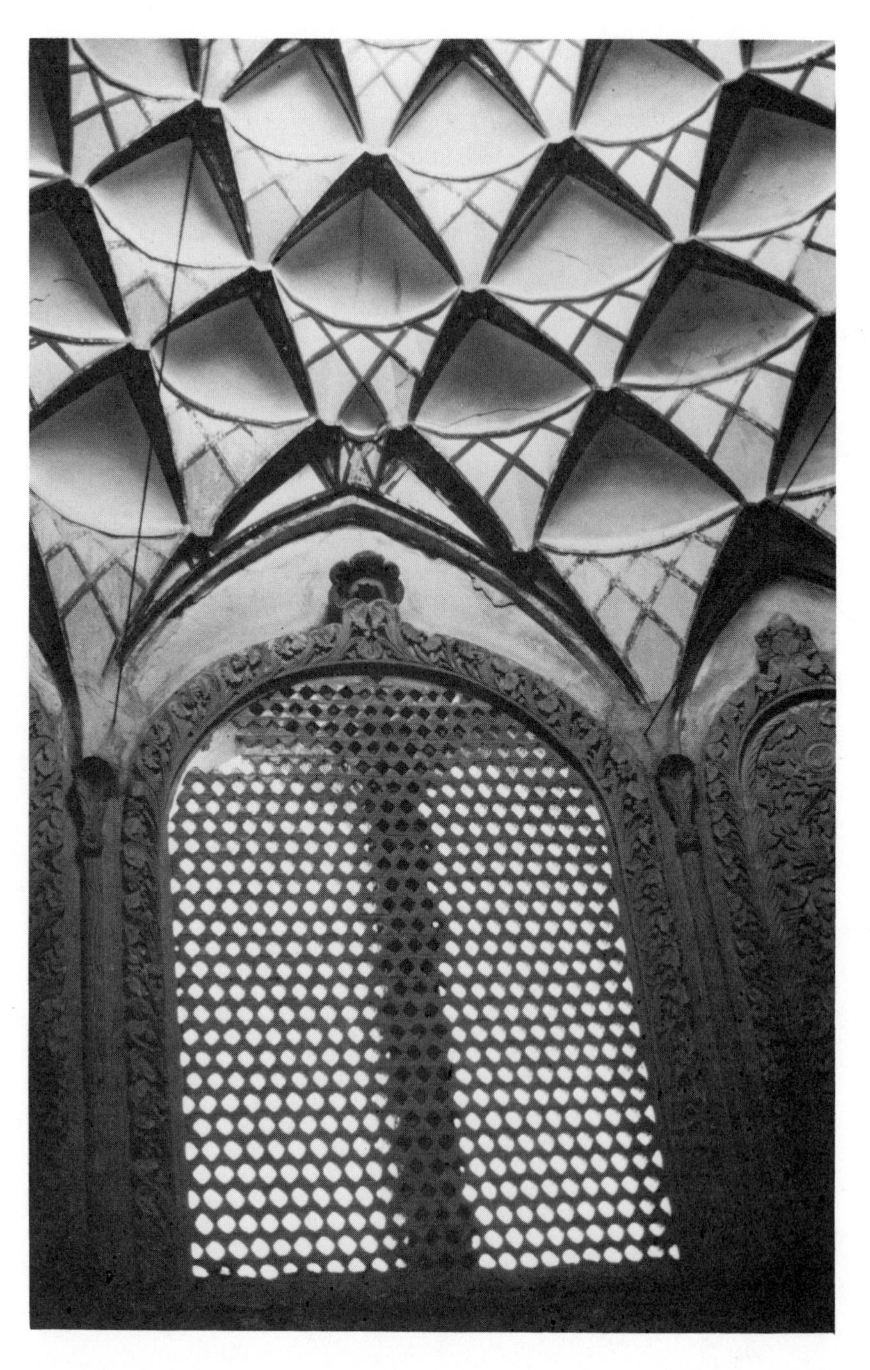

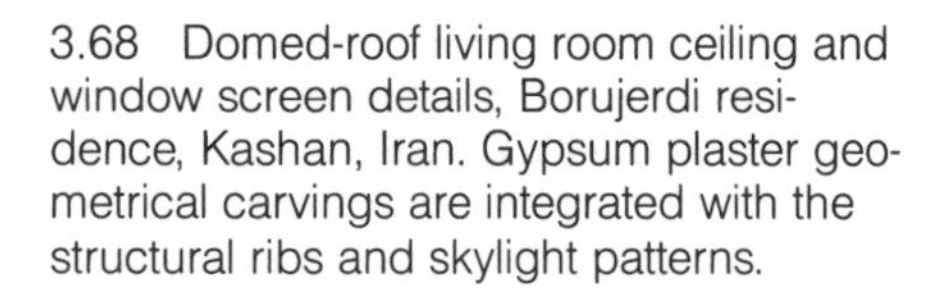

3.68 Domed-roof living room ceiling and window screen details, Borujerdi residence, Kashan, Iran. Gypsum plaster geometrical carvings are integrated with the structural ribs and skylight patterns.

sewage connection, with a pipe as a vent to the roof, can usually save an environment from being polluted.

The common rule for plumbing and sewage in an adobe structure is to take water away from the building. The cesspool or sewage connection must be far enough from the building so that the water will not run back toward it and seep up the walls.

The plumbing and electrical conduits may either be put in the wall during construction, or located in the notched adobe joints and secured later, before the plastering is done. If plumbing is left exposed, especially in low-cost habitats that use cheap fittings, leaks must be detected and fixed before they do extensive damage.

Designing and constructing electrical, plumbing, and sewage systems for a house is not a complex job and should not be feared. Some basic skills, learned from books and friends, will allow us to be self-sufficient in these area. I do not mean to exclude professional electricians and plumbers from their jobs. But in many parts of the world finding such skilled people, and then being able to afford them, is a difficult prospect that could discourage anyone from building a basic shelter. If plumbers and electricians, like architects, engineers, and bankers, cannot be done without, then the people of this world can never hope for a shelter.

Fired structures, which basically become solid masonry buildings, can use the same electrical, plumbing, and sewage system as common adobe or masonry buildings. But there is some interesting innovative work that can be done before firing. For instance, if we put wood strips in the wall or ceilings wherever a conduit is needed, the wood strip will burn out during the fire and create a chase to receive the electrical conduit or pipe. Steel pipes may also be put inside the walls before firing; since the firing temperature is only around 1,000°C (1,830°F), steel pipe (which has a much higher melting point) will not be damaged. The joints between the adobes may also be left recessed or notched before firing. After firing, the utility lines (which will be covered with the finish plaster) can be installed. Larger notches or channels must be cut before firing when the adobe is soft—otherwise, to cut hard fired brick, we will need a chisel and hammer! The use of lead-covered wires and "plaster wires," where codes allow, can safely be used.

3.69 Private residence, Kashan, Iran. The organic mud-straw roofs and the contrasting intricate craftsmanship of the gypsum plaster interiors are typical of this desert town architecture.

Earth, air, fire, and
water are obedient creatures,
they are dead to you and me,
but alive at God's presence.
—Rumi

P·A·R·T **4**

Firing and Ceramic Glazing

C·H·A·P·T·E·R 17

GELTAFTAN: FIRING EARTH STRUCTURES

bâd-o khâk-o âb-o âtash bandehand
bâ man-o tow mordeh, bâ haq zendehand
Wind and earth and water and fire are obedient creatures,
they are dead to you and me, but alive at God's presence.
—Rumi

Ten years ago I wrote that "The simple elements of earth, water, air, and fire can still create, if the magic of their intimacy is understood, the most perpetual relationship between matter and spirit." Today I am even more convinced that the most suitable environment can be created, and the greatest ecological balance can be achieved, when these elements are in equilibrium. I believe this not only because of my search into the philosophical thinking that extends back to classical Greece, where Thales spoke of water as unifying matter; or to Persia, where physician and philosopher Avicenna treated the human body based on the equilibrium of the four elements; but also because of very technical, twentieth-century facts.

There is no room in this book to talk about the philosophical or spiritual sides of the unity of the four elements, nor have I the depth of knowledge to do justice to the subject. But my experience of the microcosm of earth and fire in earth architecture, and my humble reaching for the thoughts of mystic masters, such as Rumi, have led me to believe in the macrocosm of the balance of the universal elements. But here we must talk only about the tangible and so-called physical reality—of the medium of fire, not the magic of fire.

Once fire is introduced to adobe and clay, it changes the characteristics of the earth mixture so radically that its most vulnerable point—disintegration in water—will change to its strongest point—permanent resistance to water. And that is the difference between a piece of sun-dried adobe (three elements), and a fired adobe (four elements). The missing link, fire (heat), moves earth architecture towards its perfection.

Fire introduced to adobe and clay buildings can not only create a more durable structure; it can also be utilized as a cleansing agent, to get rid of vermin or disease without using environmentally damaging poison sprayers. Fire in adobe and clay buildings can permanently sculpt interiors, and can provide a finish with the eternal beauty of ceramic glazes.

Once the fire is brought to the building, instead of building materials being taken to the fire, a completely new set of possibilities is created. One such possibility is that the building can become a producer of material instead of only a consumer. While firing a room, we can use the space as a kiln to bake bricks, tiles, pottery, or even the household pots, and dishes.

The field is wide open, and with vision and effort the unlimited possibilities can be explored by architects, ceramists, sculptors, interior designers, artists, landscape architects, scientists, and above all ordinary people, who like to put their hands into the good earth. And this may open up new hopes for the poor of the world to acquire safe and beautiful shelters with the only material available to them—earth, water, air, and fire.

WHAT IS A KILN?

A kiln is a room, small or large, where bricks or pottery or other products are baked. We must think of a room or even an entire building as a kiln. The best experience we can get is to start making some adobe and clay kilns. But the main purpose is to build and bake the kiln itself in the most efficient way.

It sounds simple, and it is simple. But we must first understand the basic elements before we can successfully fire a room.

HOW TO FIRE A ROOM

First provide the necessary flues and fill the room with whatever we want to fire. Then position the burners. Block the door and window openings with adobe, stacked without any mortar. Then put a thin mud plaster on the outside to close the open joints so that the fire cannot escape. We can do the plastering a little later, while the fire is burning. The simplest way to stack adobe blocks in the opening is to stand them next to each other and fill the corners and top with small pieces. Light the burners, and fire the room until it is done.

Updraft Circulation

If we build a wall in front of an open fireplace, we will have an updraft kiln. This means the fire is started down below, and the flame and hot gases move up through the flue and out the chimney. So if we build a small room, create a fire in front of its door and provide a hole on the top of the roof, we will have an updraft circulation, or an updraft kiln.

Native Americans of the pueblos and many other world cultures make their bread ovens this way; and the updraft kilns are used by all nations of the world. The traditional technique of brick making in both Iran and Mexico, which does not use a kiln structure, is also very similar.

In Iran, the fire is started in an underground tunnel. The heat circulates upward through the holes above the tunnel and through the stacked, sun-dried adobes. The outer layer of adobe is arranged with a flat face, called a "mirror" form. They also put a 1-centimeter- (⅜-inch-) layer of mud-straw plaster over the adobe blocks. This mud-straw layer becomes the kiln. The stack tower is open on top to allow the water vapor, fire, and hot gases to escape. The fire penetrates through the adobe joints all the way to the back of the mud-straw plaster and bakes it to brick.

The Mexican system is similar, except that they make smaller tunnels in the stack over the ground, fill the joints of the flat face layer of bricks

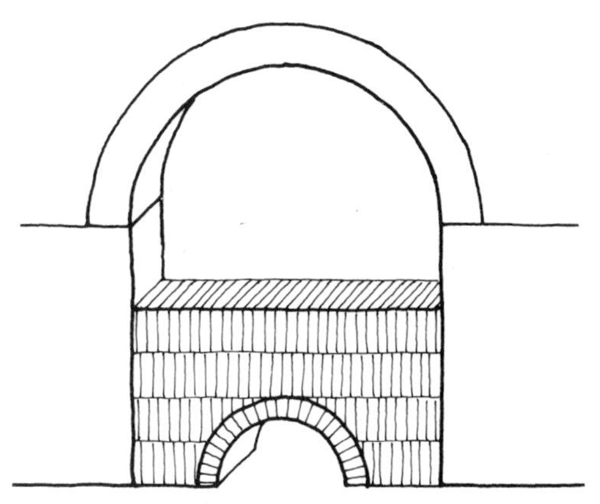

4.1 A burner opening is left at the bottom and blocks are stacked without mortar to make a temporary wall in a door or window opening.

4.2 Traditional brick firing (without a kiln structure) in Iran. The adobe blocks in the foreground are fired, while the second stack in the background is ready for the flat-face brick layer and mud-straw plaster put on before firing. An underground tunnel is used for the firing.

with mortar, and do not cover it with additional mud-straw plaster. They also use mostly animal dung for fuel. This form of low heat firing and the shorter firing period is the reason that Mexican architects and engineers are not satisfied with the bricks produced, which are not fired as well as they should be.

There are some disadvantages to updraft circulation firing, especially in a kiln. A lot of the heat is wasted, because it is pulled directly outside. Also, it does not provide uniform heat to all the spaces within the kiln. To remedy these and other problems, the downdraft circulation system is used.

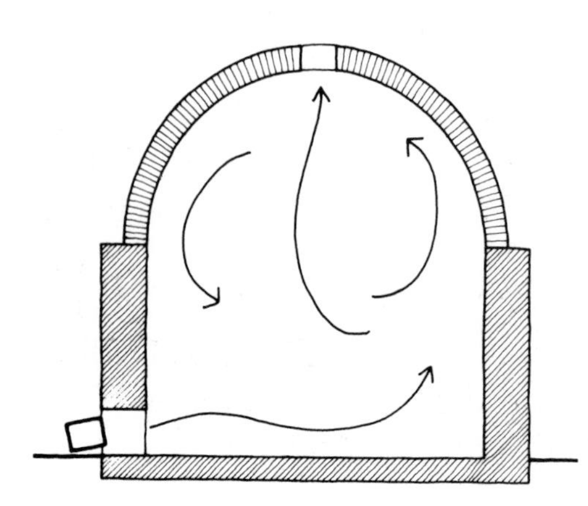

4.3 Updraft circulation.

Downdraft Circulation

The firing system in a downdraft kiln is similar to the updraft, except that instead of leaving a hole at the top of the kiln (or the room) for the fire to escape, the holes (which lead to the flue) are at the bottom.

When the fire is started, the heat circulates up to the ceiling and around the room, and comes down again and to the holes at the floor level or at the base of the walls to go out. This way the heat is kept inside much longer, and flues lead it to different sections of the room.

Flues may be provided in the walls, or a temporary flue may be provided at the center, to be removed after firing. However, perimeter flues give a more satisfactory result. The fire in this system may be introduced from a tunnel under the room, from door or window openings on the sides, or even from above the roof.

In the first few hours of firing, a tremendous amount of vapor builds up in the room, making it difficult to vent out in a downdraft system. It is desirable to exhaust these vapors as quickly as possible, to prevent cracks. To help alleviate this problem, provide some holes on the roof for the updraft circulation in the first several hours; then, when all vapors are out, close the holes by putting adobe blocks over them. Thus to fire and bake the adobe and clay structure, we are actually using a combination of updraft and downdraft circulation for the first several hours, and then downdraft circulation for the rest of the firing time.

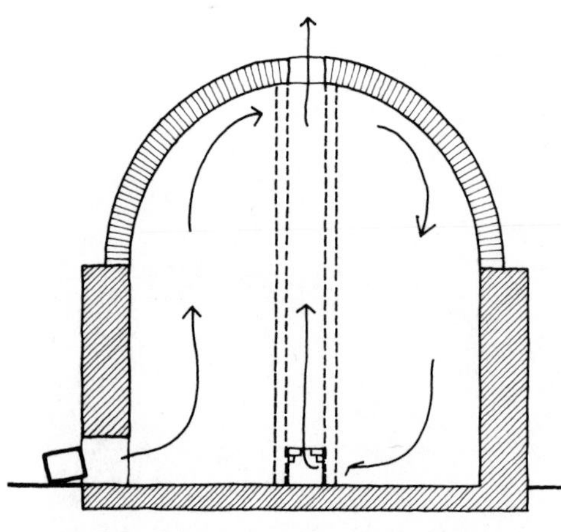

4.4 Downdraft circulation with temporary interior flue.

Updraft and Downdraft Circulation

In this combination system the room must have vent holes at the top of the roof, and flue openings at the floor or below the floor level. If we start with a simple square room covered with a dome, say 3 × 3 × 3 meters (10 × 10 × 10 feet), we can build and fire it as follows:

1. Leave a hole on top of the dome a little smaller than the size of one adobe, about 20 × 20 square centimeters (8 × 8 square inches).

2. Provide four flues, one on each wall. This is done during wall construction by leaving out one adobe at the center of the wall at each row, from the base of the wall all the way to the top. These flues can end at

the roof level. The curved roof and its valleys will work like a chimney in both domes and vaulted rows of rooms.

3. After several hours of firing at a very low temperature, most of the vapors will escape through the hole on the roof.

4. Close the top access with an adobe block and mud. Now the firing circulation changes, and heat circulates along the walls and ceiling and the hot gases escape from the flues starting at the base of the wall.

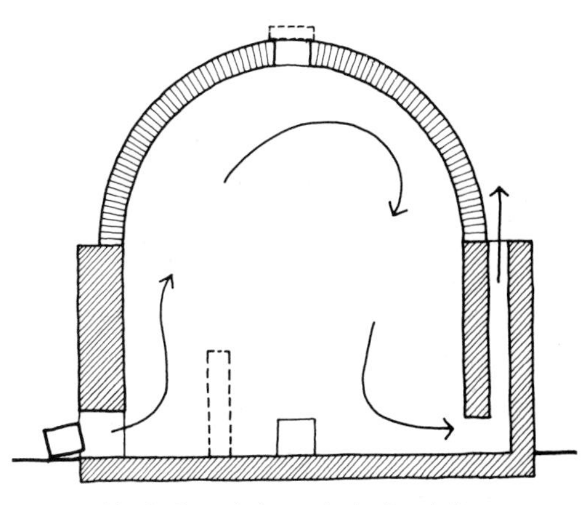

4.5 Updraft and downdraft circulation with flues in the walls.

In this way, the fire bakes the ceiling, walls, and floor. If the flue openings are located below the floor or lower, the foundation can also be fired. The upper access on the roof could be incorporated into the design as a skylight, and the flues could at last be incorporated as vent shafts for a heater or wind catcher. At the desert village, a farmer uses one of the shafts in the wall for storing his dairy products to keep them cool, as there is no electricity for refrigeration.

The flues do not necessarily need to be part of the wall; they could be removed after the firing process. However, if flues are incorporated as a part of the structure, the walls will hold more heat and use fuel more efficiently.

Crossdraft circulation, in combination with updraft, may also be utilized in firing the structures. Crossdraft is the system of locating the burner on one side and the flues on the opposite side of the room, or kiln.

FLUES

A flue is a shaft that acts as a chimney for the passage of hot air, vapor, gas, smoke, and so on. The size, location, and number of flues depend on the form and size of the space. The general rule followed by traditional kiln makers is that there should be at least one flue in every wall; in long walls, there should be one flue at least every 2 meters (6 feet). They must be large enough to draw the water vapor, hot gases, and smoke easily. Concentration of too much vapor inside of a room, without an easy escape route, will cause firing failure, surface cracking, rupture, or may even cause a piece of wall or roof to blow out.

The rule of thumb is to provide one flue for each wall, with a minimum area of 20 square centimeters (8 square inches), for a square room of 3 × 3 meters (10 × 10 feet). More flues are added as the room becomes bigger. If a flue opening is too large, or there are more flues than necessary, they can easily be adjusted by covering the holes on the roof with adobe blocks. But if the flues are too few and their openings too small, then the firing will be inefficient and no adjustment can be made during the firing. An improper circulation often causes the kiln to "choke," putting out the fire. The supply of air to the fire is the most important factor in keeping it going. The air is usually taken in at the point of the fire, depending on the system of the fuel and the burner.

The size and number of the flues are main factors in creating the draft necessary to bring in the air to the fire. In conditions where chimneys are used to vent connected flues, the height of the chimneys become important. In firing a room, the flues are individually open to the outside above the roof and a separate chimney is not needed.

It is invaluable to consult with a potter familiar with kilns, or a kiln operator, before firing, especially in different parts of the world with different fuels and traditions of firing. Even though firing technique can be learned, still it must be remembered that firing is a great art, and the kiln operators are the artists who know the behavior of fire by the way it lights, continues to burn, and dies. Knowing how fire grows hotter and recognizing how the different colors of the fire represent different temperatures are part of the art of firing. Even the most skillful of the kiln makers are sometimes surprised by the fire's behavior, and how well a kiln fires. Ali Aga, the kiln operator from Hamedan, could tell the temperature of fire by looking at it. He would even talk to fire on occasion.

WHAT HAPPENS DURING THE FIRING?

No matter how old or dry adobe and clay buildings are, even centuries old, they still contain more than 15 percent moisture. This creates a lot of vapor, which will try to escape during the early stages of firing – around 200°C to 500°C (400°F to 900°F).

As the moisture leaves the adobe or mortar, the materials soften – becoming almost like wet clay again – and the structure is at its weakest point. In Iran we had an experienced kiln operator to tell us the strength of the roof and the structure as the firing proceeded. So we were able to stretch out on top of the roof to watch the stars, while the roof was hot. But until all necessary tests are made in the future, and the degree of the temperature and the corresponding strength of the roof is determined, we should avoid the roof as long as the fire is burning. Walking on top of the thick walls or between domes and vaults, however, may be safe.

It is important to get the building as dry as possible before introducing a higher temperature fire. Start a low fire at first. If two or more burners are used, start only one of them for the first few hours to get the building dried out. Then light the others.

During the early firing, all flues and access holes on top of the roof should remain open. It may even be better to leave the stack of adobe in front of the door or window unplastered for few hours to help the early steam escape. Using mud-straw plaster during the fire is possible. The idea is to let the adobe and clay dry outside and inside, and let water that is trapped between its molecules escape as easily as possible.

After ten to fourteen hours, when the steam is all out, close the top access holes. Now the system of updraft and downdraft circulation changes to downdraft. This means that the only way fire and hot gases can escape is from the flues at the base of the wall, or below the floor. Flues at the top must be open and clear of any obstructions.

After all the steam is out, we can raise the fire to its highest temperature and let the heat build up. To fire a room and bake it to brick, of course, will depend on many factors. The most important is the amount of heat and the number of hours of firing. And if we want further solidification and fusing, we could actually melt the surfaces to create rocks. For an ordinary adobe and clay building to reach the stage of fired brick, a temperature of 1,000°C (1,830°F) is usually sufficient, depending on the clay and the depth of desired firing. To reach this stage of firing within an average size room and using a gravity-flow kerosene oil burner takes from twenty-four to thirty hours. If the room is next to a room that has already been fired, it will take less time because the walls will be much drier.

SAFETY RULES

One of the greatest advantages of adobe and clay buildings is their fire resistance. If all buildings were built of earth, fire insurance companies would go out of business. Fire may be our best friend or our worst enemy among the four elements, depending on the amount of our understanding and care.

Firing adobe and clay buildings must not present any fire hazard if it is done *correctly,* and by experienced people, since the fire is well contained. The precautions for firing structures are similar to those taken when firing adobe kilns in the open air. Depending on the type of the fuel and the system of firing, the most care should go into the type of burners and supply lines used.

A fire caused by the accidental tipping over of an oil barrel on the roof, or leakage, need not create panic as long as the building is not near other flammable objects. In one of our firings, a heavy storm caused one of the oil barrels to fall on the adjacent room, starting a large fire. We simply stood far away and watched the fire until it was burned out, since it could do no damage to our adobe and clay building. But our building was large and isolated and there was no danger of neighboring houses catching on fire.

The biggest fire hazard is probably crack, rupture, or collapse of the building during the firing. Just as a conventional building will collapse if there is not enough support in the structural members, an adobe and clay building roof may collapse without or with firing if there is not enough strength in the walls or roof. We have not yet experienced a roof collapse. However, because the structure will expand and contract during firing, we must take all possible precautions. It is of the utmost importance to have someone experienced in working with kilns to help and advise us.

The shut-off valve for oil and gas burners should be far away from the burners. If the burner or the line catches on fire, the fuel can then be shut off immediately. The connecting oil line to the burner should either be made of steel pipe or reinforced and noncombustible materials that will not expand, rupture, or catch fire if the line gets hot. Also refer to the oil burner discussion in this chapter for safety pointers.

Use a pair of dark glasses when looking through the observation hole. The heat comes out of the hole like an invisible flame, and can cause burns to the eyes and face. Anyone who has seen a ceramic kiln operate knows how intense the heat is.

When firing a building or several buildings the help and advice of an experienced fireman will be needed. At the beginning, we should plan to learn about how to handle fire, just as we learn about adobe and clay and the construction techniques. The firing of a building is an integral part of the structural system and must be learned and done correctly.

Avoid going on top of the roof during this time – the walls and the roof are still soft and may not be strong enough to support our weight. Later in the firing, when all vapors have been released, we may walk on the walls if we need to inspect the flues.

It is best to fire several rooms, or even the entire building, at once, but never get involved with the firing process if you are not in total control. The twenty-four to thirty hours of firing is enough to create a fired-shell structure; however, with slower and longer firing systems, deeper penetration of the fire and more solidification of adobe and mortar is ensured. The best handmade bricks sometimes take as long as three to five days of slow firing.

The wind can be a great annoyance and can be very dangerous. The best time to fire is during off-wind days. However, if we anticipate the wind we won't be caught off guard. If we are using oil burners, we must protect them from the wind by stacking adobe block with some sort of protection covering it for a wind break. The wind can paralyze some firing systems. A good pair of gloves and shoes – not made of rubber or plastics – are good safety measures.

The best way to get experience is to fire a small structure or room. This will help us to find the right temperature, firing time, and depth of penetration. Then we can utilize this information to fire our final structure. The best safety rule is to get help from a person who is experienced and knowledgable about firing. It will only be in the future when we have enough experience behind us, that we can standardize the techniques and safety rules of how to construct, fire, and glaze buildings, and build faultless gravity flow burners. At the present the best sources are potters, kiln operators, and the many books available about kilns.

FIRING AND FUEL

Firing and baking a room from within is similar to firing a ceramic or brick kiln. The firing system depends mainly on the availability of the fuel and local know-how. In the West and other technologically advanced countries, the varieties of fuel and systems of high-temperature firing are standard and sophisticated. Most of the equipment works on natural or propane gas, electric pumps that eject fuel, and even all-electric coil systems. But in the rest of the world, such fuel and equipment is nonexistent. Even electricity is a great luxury.

The basic philosophy of *geltaftan* is earth architecture created from the four elements—earth, water, air, and fire. It is based on human knowledge, appropriate technique, and a technology that works with gravity, the sun, and the four elements. If we extend this philosophy, it becomes obvious that sophisticated equipment should not even be considered. Not that there is evil in such high technology; but it is a far-fetched reality for most of the world, and should be left alone. High technology is appropriate for societies that have it. It is easy for Western societies to use gas and fuel-ejecting pumps to fire a room or a whole building. But here we consider only a basic system, based on locally available fuels or easily imported ones, which could be used anywhere in the world.

Even the use of imported fuels such as oil is questionable, since it may be a strain on the economy. In that case it is good to compare the available alternatives and calculate the tradeoffs. We cannot say that constructing adobe and clay buildings and firing them is appropriate for every place and every condition; this would be a foolhardy claim. *Geltaftan* is an alternative—and a more valid one than many others—that could suit many parts of the world.

Local Firing Systems

The best firing system is usually the system used locally to fire brick and ceramic kilns. The fuel can range from coal and coke to wood, grass, and weeds; from animal dung and trash to natural gas, oil, bio gas, and electricity; or even solar energy, microwave, and fusion.

A traditional Mexican fired-brick maker still uses animal dung to fire his brick kiln, even though the country is rich in oil. The oil has not reached him, and the man has no time to make the transition from animal dung to oil. In Iran most traditional village kilns are similar to that of the Mexicans, but the fuel has already been changed to oil. And even those who used animal dung have found an easy gravity system to burn oil. The oil is stored in a tank (usually a ditch dug into the ground, plastered, and used as a tank). Then the oil flows through a steel pipe into the tunnel under the adobe stack and burns.

A small tank of water may also be utilized; a pipe carries water next to the oil pipe. Drops of water and drops of oil simply flow down and splash over a piece of fired adobe in the tunnel floor and catch on fire. The water drops are very slow, one or two drops per second, while the oil comes in a continuous flow as needed. Water drops splash and break, and the oxygen in the water helps the oil to burn much better. As the heat rises, the smoke and vapor go through the bricks and up and out. The mud-straw plaster around the stack of unfired block is the kiln, and as the kiln gets hotter the black smoke created by the oil burner is burned out. In this system there is no chimney stack or flue. And the fire is a beautiful fire, sometimes resembling the fireworks of celebration nights.

There are many variations on this basic system. There may be ready-made adobe kilns on top of the firing tunnels, instead of the mud-straw plastering around the brick stack. There may be a steam line instead of liquid water drops (the heat of the kiln makes the steam). If power is available, a small electric pump can change the oil and water to moist spray and eject it into the room. We can utilize any of these systems for firing an adobe and clay building, both for baking it to brick or glazing it for ceramic finishes. A local potter or kiln operator can give invaluable information. Of course, their word is not the last word, since our kiln is bigger than theirs and we need to reach out for all other possibilities.

A Gravity-Flow Oil Burner Firing System

Since oil is available in most countries of the world, and diesel and kerosene can be acquired in more places than other fuels, we will discuss a homemade burner system based on this idea. This is not the only system everyone should use, but we have found it to be simpler than most.

The operating principle is based on the flow of oil via gravity into a burner, which changes the liquid fuel naturally to gas and fires the kiln or the room. It works like this:

1. Put one or two (connected) barrels of oil on a roof or tower approximately 6 meters (20 feet) high, and connect the fuel line to a burner so that the oil flows into the burner.

2. Start a fire in the burner. It will burn as long as there is fuel in the barrels. A valve controls the amount of oil allowed to go to the burner.

Placement of the burners is an important consideration. One or more burners in a room or several rooms can be located below the floor, or in the door and window openings. A square room of approximately 3 ×

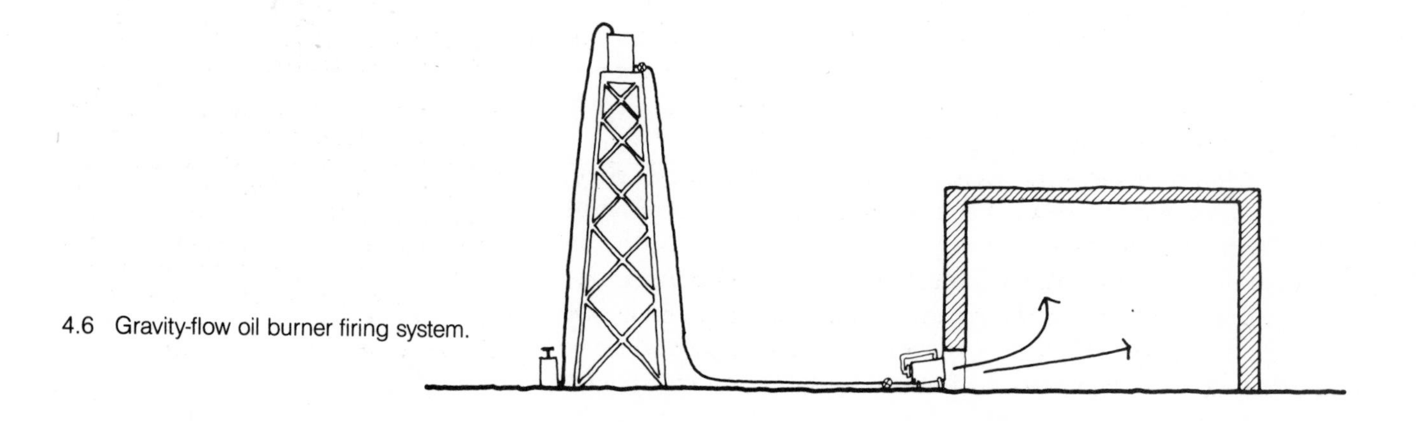

4.6 Gravity-flow oil burner firing system.

3 × 3 meters (10 × 10 × 10 feet) with a dome can be evenly fired with a single homemade burner located under the center of the room.

Such a room can also be used as an ideal kiln for pottery firing. Almost all traditional ceramic kilns in Hamedan, a city in Iran, use this system to fire their vessels. But digging a tunnel under a room can be hard work if we are firing a whole building. In this case, the system of side firing may be utilized. Locating the burner in front of the door or a window opening is the simplest way to fire a room. The fire inside the room is directed for even distribution, depending on the location of the flue. In a room with a door and a window on opposite sides, two oil burners can do the job well. This is used most commonly in firing of a vaulted-roof, rectangular room.

Firing from above, as is done in some brick-manufacturing kilns, can be adapted to the firing of the buildings, if the project is large and the cost of the system can be justified.

Hamedan Burner: A Homemade Oil Burner

One of the best kerosene oil burners we have found is the **Hamedan burner**. (Hamedan is an ancient Iranian town, famous for its ceramics.) It works without any need for electricity, pump, or other gadgets. This gravity burner can be constructed and operated by anyone, after gaining some experience. We have adapted the original design to suit our work, but have kept its simplicity.

The burner works like this: Elevate a barrel of kerosene oil to about 6 meters (20 feet) and connect it with a pipe to the burner. The oil flow goes through the hot burner, changes to gas, and shoots flame into the room. The burner is heated for a few minutes at the beginning, but works from then on with its own heat. The oil flow is controlled by two valves—one at the high point where the tank or barrel is located, and the other near the burner. It is the second valve which is adjusted; the first valve is simply opened at the beginning and shut at the end of operation.

This burner may be made larger or smaller, but an average-size burner with ample fire, and light enough for two men to move around, is detailed and described here. This burner would be sufficient to fire a pottery kiln or a room of around 3 × 3 × 3 meters (10 × 10 × 10 feet). In full operation, with the valve completely open, it uses an average of around 32 liters (8 gallons) of kerosene oil per hour.

The Hamedan burner works well with kerosene oil, although diesel oil may also be used. However, it is less efficient and has a shorter life because of residue build-up.

The burner is made by putting a small steel pipe inside a larger one, and welding them together at the top and bottom. The air space between the two pipes will be filled with kerosene oil during operation.

The step-by-step operation is as follows:

4.7 Two oil barrels on a tall stand are connected together at their base with a pipe to keep the oil at the same level in each. The oil flows out through two flexible hoses down to the burners below. Another hose is connected to fill the barrels with a hand pump on the ground. The ladder is used to monitor the oil level. (Instead of a tall stand on the ground, a rooftop could be used to gain the needed height.)

1. First open the valve at the tank, and then open the control valve near the burner. As soon as the oil starts to shoot out of the burner, shut the control valve.

2. With a few sticks, start a little fire around the burner. If the burner

4.8 A burner located in front of a door opening to shoot the fire inside. (The temporary wall closing the opening was not constructed here so that the burner's fire in the first several minutes can be seen.)

4.9 The burner's fire after it catches on with a roaring sound. The volume of flames decreases but the intensity of the fire increases.

is vertical, start the fire underneath; if the burner is horizontal, start the fire inside.

3. When the burner is hot, try to open and shut the valve a few times until the roaring sound of fire is heard; then adjust. (The oil is changing to gas and shooting through the burner by itself during this period.) If we open the control valve and oil shoots out, it means that the burner is not hot enough.

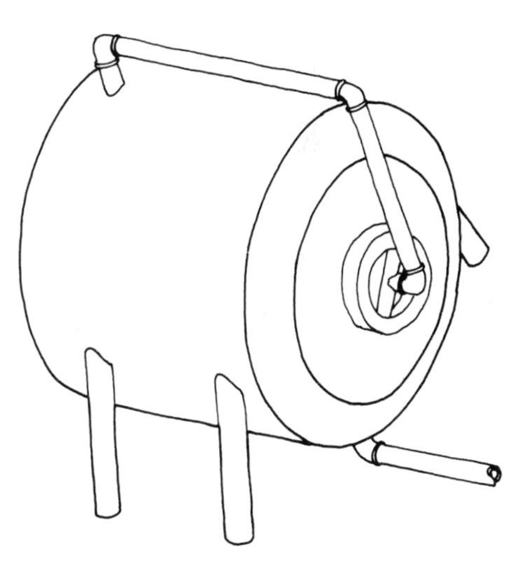

4.10 Kerosene oil burner.

Remember that the fire must be kept very low the first several hours. The control valve must be adjusted accordingly. The fire will catch on and continue to burn, not inside the burner, but inside the room.

The burner should continue with no problem as long as there is oil in the tank and the supply line. The oil flow keeps the pipelines cool and will not let the heat get into the main line. It is better to keep the oil on top of the tower at a regular temperature and shade it from the excessive heat of the summer sun.

The air supply for the burner is from its open end. The ejecting gas takes the air with it to be fired in the room. Thus it is important that the burner is located in an area where it can take as much air as it needs. If a room or a kiln is fired from underneath, the tunnel should be wide enough for both the supply of enough air and for someone to crawl to the burner. When the burner head orifices get clogged with dirt, a wire pin attached to a handle can be used to clear them.

The gravity flow firing system is almost failure-proof, since it doesn't depend on any mechanical equipment. It can be used over and over again, anywhere in the world. Just remember a few main points: The burner must be located completely outside of the room and must have an air supply around it. It must be protected from the wind. It must not get overheated, which will cause it to die. (Sometimes overheating is caused by not having a good flow of air or oil; or oil that is hot because of exposure to too much sun.) The burner generates a very powerful fire and can be dangerous if it is not handled with care and skill.

The main shortcoming of the burner, in its present design, is the lack of an emergency shut-off system. Once we shut the valve it may take over fifteen minutes before the oil inside the burner is finished and fire goes off. This burner must only be used by an experienced operator. An emergency shut-off system must be developed for it before its public use.

To supply the oil to the tank on top, we can use a hand pump. To check the fuel, climb to the tank or use a float or other shut-off system. A small compressor can also provide the needed pressure if the tank is located on the ground.

A group of domes or vaults may be fired all at once, with several burners, or one at a time. In this case, by connecting the rooms to each other, the building will become like a multiple kiln and the heat will travel from one space to the other. In sloping sites the firing is carried out from the lowest room first – the way the Japanese chamber kilns work. In this manner, the extra heat generated from one room will cause the others to dry out; and in return, less heat and time will be needed to fire the other rooms.

The room space, when used as a kiln to bake adobes, tiles, or pottery, must be utilized like local kilns. Stacked brick or jars or shelving can hold different size pots.

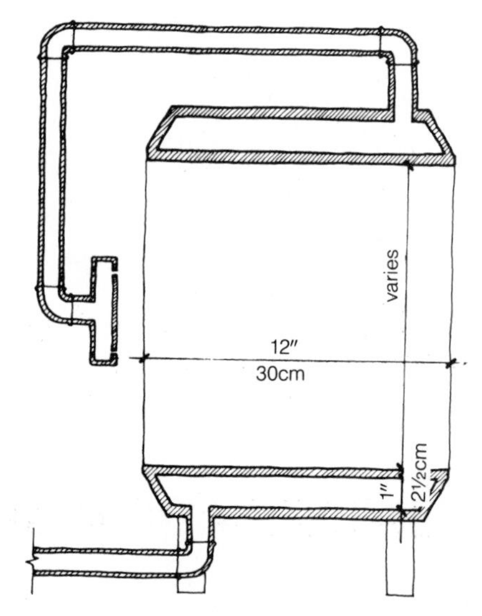

4.11 Cross section of a kerosene burner.

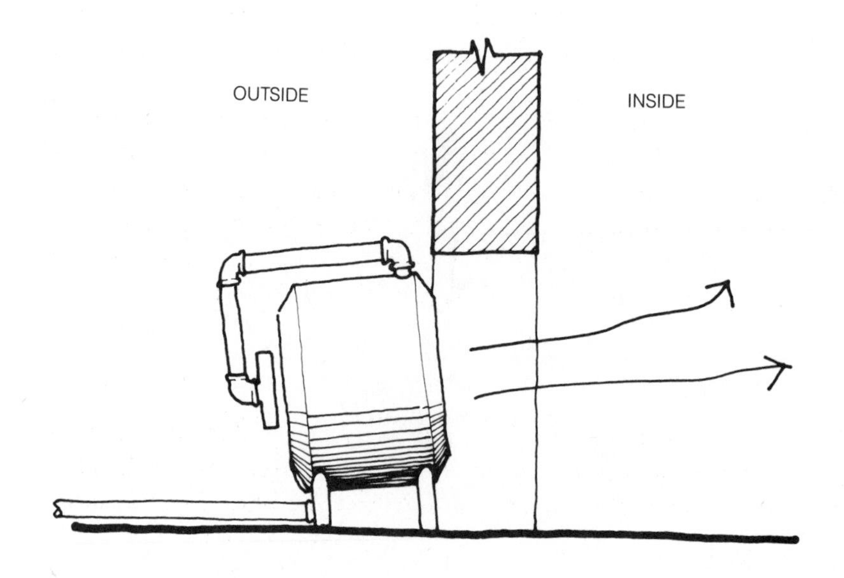

4.12 Placement of the burner. To construct a burner:

1. Cut two pieces of steel pipe of 30 and 38 centimeters inside diameter (12 and 15 inches). Smaller burners can be made by keeping an approximate dimension ratio.
2. Drill 2 holes for 2.5-centimeter diameter (1-inch) pipes in the bigger pipe. (At top and bottom.)
3. Set the smaller pipe inside the larger one and weld with a strip all around to make an air-tight chamber of about 3 centimeters (1 inch).
4. Weld black pipes with standard detachable elbows as shown.
5. Connect four legs, with front legs taller, to point burner up.
6. Provide a control valve and a check valve 3 meters (10 feet) away from the burner.
7. Make a circular burner head with twelve holes 1.0 to 1.5 millimeter in diameter.

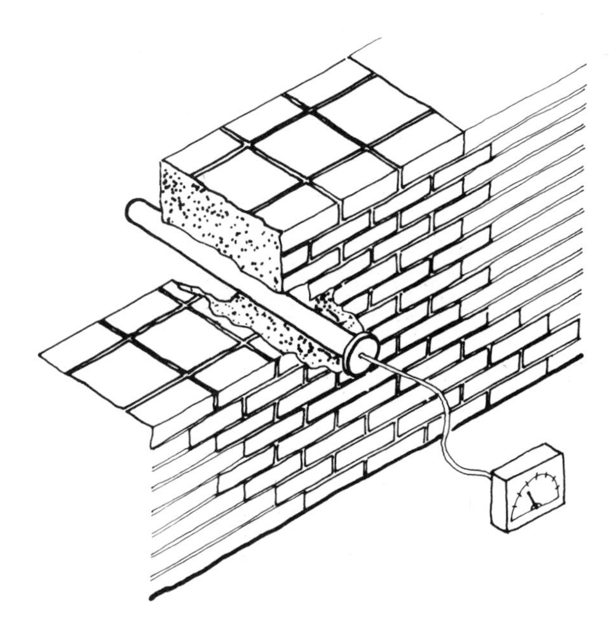

4.13 Thermocouple, to measure kiln temperature.

MEASURING TEMPERATURE

During firing, the temperature inside the room can be measured in different ways. The experienced kiln operator can tell the temperature of the fire by its color: red colors, ranging from light to dark, are around 500°C to 800°C (900°F to 1,500°F); orange to yellow to white fires are from 800°C to 1,000°C (1,500°F to 1,800°F) and above.

We can also use a thermocouple to tell how hot the fire is. It can be left in the observation hole of a wall of the room for the entire baking time, with an outside indicator to show the temperature rise. Or it could be inserted in different parts of the room and read through observation holes.

It is also possible to use the sophisticated scopes and scanners. But since the firing of a building is not as critical as the firing of a ceramic kiln, simpler methods, such as the use of cones or tile samples are sufficient.

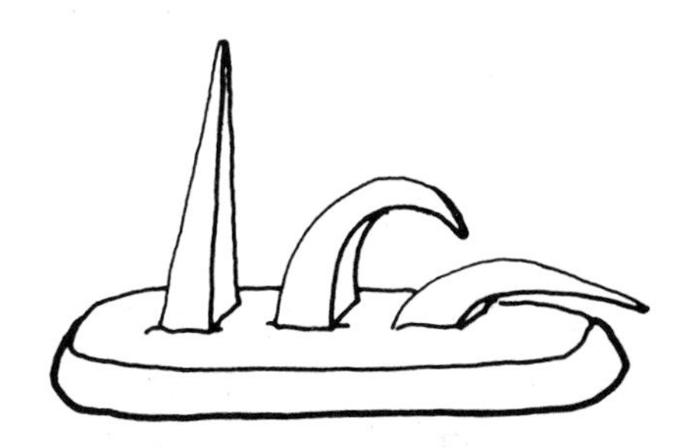

4.14 Cones, used to measure kiln temperature.

Cones are finger-size, ready-made clay pieces that melt at a predetermined temperature. Several cones with various melting points can be placed in the room, and watched during the firing through observation holes. As the cones melt, the degree of temperature inside the room can be determined.

The village kiln operators sometimes leave some bricks or tiles in the kiln, within reach. During the firing, they are taken out with a long steel clip and tested for strength. How we choose to measure the temperature will depend on local conditions and availability of materials. Ali Aga, our kiln operator, would go up over the wall, look at the flue shafts, then tell us: "When the flues are all white, the room is baked."

FINAL STAGES OF FIRING

When the fire reaches 1,000°C (1,830°F) – **cone O6** – let the fire continue for about two hours or so before turning it off. This temperature is generally accepted to be hot enough to change adobe to brick. But in reality–

depending on the adobe mixture, the thickness of the block, and other factors—it changes. Thus the desired temperature must be determined during sample testing. At this stage the mortar between the adobes inside the room is glowing. It looks like gold nuggets and shining lights; the adobe blocks are glowing as well, but in a different shade. The sight is breathtaking, and every little detail can be seen. The high temperature makes the room appear much smaller than it really is and the far end wall will look much closer.

At the last minute, shut off the valves. Then stand and watch:

> I see one of the most beautiful sights I have ever seen in my life: the dying of a high-fired space. Since the torches are off, there is no fuel and there is no flame, but the room itself is emitting fire. The walls seem to be shooting fireballs at each other. A fired room doesn't die all at once; it plays a lot before its death. The whole room after a while starts glowing in deep red color, then adobe blocks of the walls and the ceiling become amber color, while the mortar lines between them change to pure gold nugget. I feel that there is a lot of fire inside those masses even though I don't see the flame.

Then comes the cooling period. The only openings should be flues on the roof, through which the air and heat exchange occurs. Through these shafts, and through the roof and wall surfaces, the room cools off very slowly. Let it sit for forty-eight hours or more before opening it up. Early opening creates cracks.

After the firing is finished, the burner can be removed and used somewhere else and its portholes (the openings in the wall) can be closed with adobe and mud. When the room is cooled and opened, the simplest and yet the most profound sight can be seen and touched: The entire adobe and clay room has changed to solid brick. What used to be adobe blocks and mortar, clay-earth rammed construction, or piled-mud wall—structures that could have been washed away by water—now has become water resistant. The dried-out mud, which we could once have broken with our fingers, now requires us to use a hammer and chisel.

C·H·A·P·T·E·R 18
INTERIOR DESIGN: SCULPTING AND GLAZING

Earth, enchanted by fire, creates dreams in forms, colors, and texture as vast as the cosmos.

Clay-earth is an ideal medium for sculpting and creating spaces and molding the imagination. The interior spaces of a fired adobe structure can be designed, sculpted, fired, and glazed to dream shapes. To build a room with adobe and clay, and then to fire it and leave it alone, is itself a rewarding experience. But to sculpt and glaze the interiors – including textures, paintings, graphics, shelves, fireplace, and even coat hangers and furniture – is to integrate art into architecture. The artists are as much a part of this architecture as the architects are part of their arts. A potter becomes an architect, a sculptor becomes an interior designer, a painter becomes a ceramist, and an architect becomes a sculptor.

While under construction, adobe and clay structures can be sculpted to attain an aesthetic spirit and then be frozen in place by fire. The addition of small amounts of oxides or different types of clay and sand to the blocks can create fine finishes when fired. Adobe and clay walls made thicker than is structurally necessary may be carved and textured. Locally available gypsum or cement plaster or even whitewash can be applied to create contrasts and light-reflecting surfaces.

Ordinary clay-earth, gypsum, and straw, used in a method called *sim-kah-gel* in Persian, can result in some of the most beautiful and lasting interior finishes. First, a band of gypsum (white plaster) is applied around doors, windows, and other designed edges called screeds. Then we apply the buff color. This plaster is a mixture of clay-earth, straw dust (very fine, screened straw), and water, which must sit until yellowish water ("juice") appears in the container. Then the mixture is well kneaded to a paste. After this plaster is applied and before it is dried, take some of the juice from the container and cover the screed white plaster bands with it. Then, before it is completely dried out, take a putty knife or similar tool and scrape the half-dried juice from the top of the white screeds. The white plaster color will come through the buff layer to create a homogeneous, decorative surface that looks inlaid. Here again, the main material is the earth, and the white plaster is used as the decorative element.

The interiors of earth architecture buildings constructed with arches, vaults, and domes will look beautiful if the sun is allowed to play on the surfaces. All other treatments are an opportunity for the artists to interact with the earth and the sun.

CERAMIC GLAZING

> Now the time has come to create a new scale in the ceramic world, to walk out from the womb of a pot to the space of a room.
>
> Now the time has come to step back into history, and touch our fathers who have touched the glaze, to recapture their secrets of heavenly textures and colors, but grow larger than their size, to create, not only the little forms and shapes in containers, but the spaces that contain us.
>
> Now the time has come to create a ceramic glaze, a china, a stoneware, not in the scale of our hands but in the scale of our lives.

4.15 The natural colors and textures of the fired surfaces, accentuating the structural geometrics, can be left as finishings.

The words pottery, bricks, and ceramics have always meant something we can lift and carry in our hands. These words, representing the products of the clay-earth, have limited our imaginations not in forms, textures, or colors, but in scale. And even though humans have fired and baked giant-size kilns, they only meant to fire its mini-size contents. But now, if we can free ourselves from that limitation of scale and believe that we can take fire to clay just as we can take clay to fire, new horizons will open to us. We have inherited a great wealth of forms, colors, and textures developed by architects, potters, and sculptors throughout history—and new ones arise everyday. If we could learn to use all this wealth in a new direction and new scale, then an era of beauty at the service of the human soul will emerge.

Of course, the change of scale may also present technical and socioeconomic limitations; but these limitations can be overcome by imagination and craftsmanship. The cost of fuel for fire, or the expense of ceramic glazes may not seem to be within our means; but if we analyze the situation carefully, we may find that they are appropriate.

For example, if we consider buying fired brick instead of firing the structure in place, we will find that the cost of fuel to manufacture and transport is many times higher. Or if we consider making large quantities of glaze with broken bottles, instead of buying it in grams or ounces, then covering large surfaces will become economical. Even learning how

4.16 Forms and spaces can be constructed on human scale, and exposed or plastered surfaces can create contrasts and textures in tune with human senses. Javadabad Elementary School, Varamin, Iran.

to use kitchen salt for beautiful salt glazing can make economic issues secondary to structural and aesthetic concerns. With fired structures we may find that there is no need for additional finishes. The surfaces themselves have beauty, especially in their imperfections. Exposed blocks with their joints designed for expansion and contraction, or hairline cracks,

4.17 Forms and surfaces can be simple; the sunlight does the rest. Javadabad Elementary School.

may be considered as part of the design. Think of Michelangelo's paintings, with their beautiful—time created—hairline-cracked surfaces.

Many potters are as familiar with glaze as they are with clay. But for the majority of people, glaze is still mystery. They have no idea what a glaze is. I had no idea myself when I tried my hand at glazing for the first time:

> It is almost impossible for me to separate in my mind the picture, the texture, the color, and the sensation of a ceramic glaze from that of a seashell. I have always imagined my glazed houses to have the color and textures, the soft and curved edges around the openings and projections, that seashells have. I can imagine a human walking, sitting, and living in the glazed spaces like one of those sea animals crawling into its shell.
>
> But now that I am faced with the actual operation of glazing, I am at a loss. What is ceramic glazing?

Since then, I have learned a lot about ceramic glazing—not enough to master it, but to know that there is no limit to its possibilities. The intuitive knowledge of the simple desert potters—even those who couldn't read or write their own names, but knew the fire by its color and its feel—the calculative secrets of the educated ceramists, and the fine touch of the masters have all made me thirstier for this magic potion called glaze. The great ceramists of China and Japan, the fine potters of the Iranian deserts, the Africans, the Arabs, the American, the Europeans—all can teach us new lessons. Books in libraries, or secret formulas written in old folded papers in all the languages of the world—these too are our pathways to the treasure of glaze and fire.

WHAT IS GLAZE?

Glaze is a phenomenon created by the intimacy of earth and fire. Before humans discovered how to glaze, nature was creating it with volcanoes and forest fires, and it continues to do so.

Glaze is a glass-like material that is used for its durability and beauty. Generally speaking, wherever intense heat is introduced next to the earth, glazed surfaces can be seen—from an atomic bomb explosion ground to a bonfire at a campground. But the mystery of glaze is like the mystery of the clay itself—the more we know, the more we see its unlimited possibilities.

Glazing materials used in pottery are powdered mineral oxides—oxide means the intermixture of the elements with oxygen. The most important of all these oxides is silica, which covers more than half of our planet. When silica is fired to the melting point and then cooled, it makes glass. But it takes a very high temperature to melt silica. So, to lower the melt-

4.18 Shelves and niches carved into traditional spaces —following the lines of the structural forces—can inspire sculptured interiors. Javadabad village, Varamin, Iran.

ing point of the silica, we mix it with other minerals (oxides) such as sodium, calcium, or potassium. Much lower-temperature oxides, like lead, are used as flux. Flux (e.g., lead) is the agent that reduces the need for high firing temperatures.

To make a glaze, we mix the ingredients with water until it can be brushed or sprayed on a clay-earth surface. When a glazed surface is baked in a fire, the different oxides begin to interact and melt into each other. In fusion they lose their own individual identity and become a uniform character.

The art of making different types of glazes, and the secret formulas to the oxides and the ways of firing them are as vast as the imagination. Before we get involved with glazing a building, we should spend some time learning about glaze finishes from potters and ceramists and observing their work. To work as a team with a potter and a ceramist who knows the fire and the glaze is invaluable.

4.19 A homemade glaze made with milled Coca-Cola bottles, oxides, and water is sprayed over walls, ceiling, and floor using an insecticide sprayer.

How to Glaze a Building

To glaze a building if it involves glazing of large surfaces economically, we must use the more abundant and less expensive glaze agents. It is generally more feasible to use a low-fired glaze, since it takes less heat and in many cases it is less expensive. The point to remember is that high-temperature glazes are more durable than low-temperature ones. But since we are not dealing with an intensive use factor—as in a cup or a plate—a high-fired glaze may not be needed. In areas of use such as a bathroom or a counter top, however, we may want to use a more durable glaze than we would use on a common vault surface. Unfortunately since the state of the art of large-scale-building glazing is still in its infancy, a set of charts and recommendations as to the use of what glaze to what purpose has yet to be developed.

Glazing an entire room, like covering a room with conventional ceramic tile, is not desirable. The room would not be able to "breathe," nor would it be acoustically suitable for a living environment. Thus it is a good idea to sculpt and glaze a few areas and leave the rest as exposed, burned adobe; or just glaze some of the walls, or individual rows of adobe brick surfaces only.

Glaze can be applied by spraying the large areas, and brushing it on the smaller sections. In rural areas, where there was no electricity for the use of mechanical tools, we used the farmers' standard insecticide sprayers. The sprayer is pressurized by a bicycle pump and can work wonders. Since a glaze solution is water-based, and can be diluted to a thin liquid, such simple tools are the most appropriate technology.

Ali Aga made a homemade glaze by grinding broken Coke bottles and adding a couple of oxides. He then applied it, within twenty-four hours of mixing, with an insecticide sprayer over the room surfaces. The decorative parts were covered with glaze applied by a brush.

Some care must be taken as far as safety is concerned. Lead oxide, which is used all over the world as part of the glazing mix, has tremendous advantages over other fluxes. But since lead is poisonous, handling it in raw form must be done with great care. Spraying the lead and breathing it is very dangerous and must be either completely avoided or done with adequate protection. There is no hazard once lead glaze is properly fired. An improperly fired lead-glaze dinner plate may cause contamination of food and raw fruits, but a glazed wall surface is much safer.

Bringing the fire to the product for glazing instead of taking the product to the fire requires a new way of thinking, which may alter the process itself. And sometimes what is a problem in conventional ceramic glazing may become an advantage when used for a building.

Salt glazing is an interesting example. Salt is one of the simplest, least expensive, and most beautiful glazes ever created. The potter simply throws some table salt into the fire. The salt becomes the glaze and covers the clay-earth surfaces to make a permanent finish. The possibilities of creating ideal textures and fine finishes are limitless. But it is not used commonly in ceramics because of some inherent disadvantages. For one thing, the salt glazes the kiln before it glazes the pottery. Another disadvantage is that it produces poisonous chlorine gas; thus the kilns must either be outdoors, or vented to take the fumes outside. But in glazing a room or a building, the first problem changes to an advantage—we *want*

4.20 Interiors can be sculpted (including shelves and coat hangers) and graphics integrated into the glazings. The first firing can be done while the room is also being utilized as a kiln filled with bricks, tiles, pottery, and clay objects. A second firing can be done later for glazing only. Single firing is also possible.

to glaze the kiln, which is our room. In addition, the building itself is an outside kiln, and thus chlorine gas is already outside. And since we only fire a room once, we are not polluting the atmosphere as a kiln does with hundreds of firing.

The fired surfaces must be free from impurities (such as lime) before they are glazed. As in pottery, such impurities will crack the glazed surfaces. In some of our experiments we sprayed the fired surfaces with water and allowed the soaked lime to pop through before covering the surfaces with glaze. The modest experimental glazing works I have been involved with are too few and too elementary to be exhibited; but my experience of glazed spaces has brought me inspirations that have become a part of my everyday life's dream. And I make a plea to the ceramists of the world to create and to experience a glazed space.

C·H·A·P·T·E·R 19 LANDSCAPING AND REHABILITATION

LANDSCAPING

Adobe and clay buildings are naturally harmonious with their surroundings. There are millions of houses and villages in the world raised from the earth itself that fit the beauty of forms, colors, and textures of their sites. Sometimes even an entire village goes undetected from the land it is built on. But the use of plants and shrubs near or inside of an unfired adobe and clay building may damage the structure, because the earth disintegrates in water. But in a fired structure, which is resistant to water, landscaping can become part of the structure. Planters can be sculpted and fired along with the building, and plants and shrubs can grow right out of a building. Thus landscaping can become integrated with the human-built environments.

The outside and inside gardens – including the fountain pool, steps, benches, and permanent furniture, pots, and planters – may be designed, sculpted, fired, and glazed on the spot. After a planter or bench or fountain is sculpted, a temporary kiln can be built over it, or a movable kiln can cover the piece to be fired and glazed. A temporary kiln may be as simple as a fire-blanket covering the sculpted piece. The possibilities of creating landscaping in small or large scale – from a fired-on-the-spot ceramic sidewalk and sculpted bench or planter to a swimming pool dug and fired and glazed in place – are limitless.

In most cases practical, simple, existing kiln techniques can be used. We can learn from the native ceramists of India who are firing in place huge sculpted forms, such as full-scale elephants or horses, by piling earth and blocks around them to work as a kiln. We can learn from the volcanos, which create the most beautiful landscape just by introducing molten earth. Once we move our fire with us as a friend, to the landscape, entirely new design possibilities emerge. And finally, nature herself teaches us the ultimate in balance and beauty by using the element of fire to create landscapes, as she uses the earth, air, and water in her spirit of unity.

REHABILITATING OLD ADOBE AND CLAY BUILDINGS

Millions of adobe buildings in the world are in danger of destruction—not by an earthquake, or flood, but by wet weather. Many of these buildings could be saved by fire. Millions of buildings in this world made of adobe or piled mud are infested with disease, mice, and vermin. They could be made hygienic by the purifying character of fire. An entire earthen village could be cleansed of disease and vermin by fire and glaze. Fire is holy in spirit. We can learn to use it to create a better living environment, including fireproof buildings.

Not all adobe and clay buildings are suitable for rehabilitation by fire. The material must have enough clay-earth in it to fuse in the fire and work as the cementing agent. Rocks and lime and organic materials in typical adobes are not suitable for high fire, since they break and disintegrate. However, buildings made with earthen material can be fired to become hygienic, since the low fire needed for this purpose will not damage the building. The temperature will wipe out bacteria, rats, and vermin that cannot be chased out of their hole by any poison sprayers.

Old adobe buildings that cannot take high fire because they have too many impurities, or burnable elements such as wood, can be protected by a layer of clay-earth adobe blocks or thick clay plaster. This new shell can be fired and glazed. In industrialized countries a layer of fire-blanket, such as those used to line the insides of kilns, can protect the combustible sections. Existing buildings and communities should be studied and analyzed to determine the economic rewards and their suitability for firing.

When an old building is found suitable for firing, the procedure is the same as for a new construction. We take a room, for example, and change its space to a kiln-like space. This could be done by digging a few flues in the walls for firing; or temporary flues could be introduced in the room to be removed after the firing is finished. The old plaster is removed to expose the adobe structure directly to the fire. The doors and windows are removed, and the openings are temporarily closed with adobe and thin plaster. The existing plaster over the roof or the outside of the wall can be removed, or grooves can be cut in them to allow the vapor to escape. The waterproofing clay plaster can be applied over the roof during the last part of the firing, as in new construction. The firing starts with a very slow and gradually rising temperature so that the heat penetration into the old building walls, floors, and roofs (if the roof is also adobe) will not create sudden expansion and contraction. The firing should take approximately twenty-four hours or more, depending on the desirable amount of fire penetration.

No matter how old an adobe and clay building is, it always has moisture inside it that should be first (and slowly) removed. This moisture is at least 14 percent to 18 percent, even in a thousand-year-old building, and can only be removed by high temperatures—200°C to 500°C (400°F to 900°F). Since the weather only gets as high as a fraction of these temperatures, even in Death Valley or the Sahara, the water is always available in the earthen material buildings. And as the old adobe and clay buildings heat up and the vapor tries to escape, then the adobe blocks, mortar, or rammed earth becomes soft again. And the vapor must be removed from inside the room as fast as possible through the flues or the vent opening on the roof. As fire removes the moisture, it also starts a crust over the surface closest to the fire, and the baking and solidifying gradually move deeper into the structure. The rest of the process is similar to that of firing new construction. (The book *Racing Alone* covers, in its second half, the process of rehabilitating old earthen buildings.)

Earth, air, fire, and
water are obedient creatures,
they are dead to you and me,
but alive at God's presence.

—Rumi

P·A·R·T 5

Beginning

CHAPTER 20 MODEL MAKING

The best way to put the knowledge gained from this book into practice is to start practicing. Find some clay and a piece of plywood and start building small models. The following pages will show step by step how to build at home or in school, before starting work on a site. As I have shared the information in this book with thousands of people in lectures and workshops, I have learned that:

- People from all walks of life—men, women, and children—are extremely interested in learning how to build with earth.
- Making small-scale models with clay is one of the best and fastest ways people will learn about earth architecture and how to build walls, arches, vaults, and domes. The traditional long training of students and apprentices produces good masons and teachers, but it will never be enough to make a dent in the world's housing shortage. For earth architecture to flourish, people must be educated—they must experience what is possible, and touch the clay with their own hands.
- Once a person has constructed a model—within hours—he or she is encouraged to learn and understand more. The knowledge thus gained will trigger quests in building with earth.
- It is possible to learn the basics of thousands of years of earth architecture within a day if it is taught in the simplest terms, and if all our senses are involved in the learning process.

We will begin practicing by constructing a room and covering it with a simple dome or a vault. We will use a small scale: 5 centimeters = 1 meter (¾ inches = 1 foot). Thus our 20 × 20 × 5-centimeter (8 × 8 × 2-inch) adobes become 1 centimeter or ½ inch square.

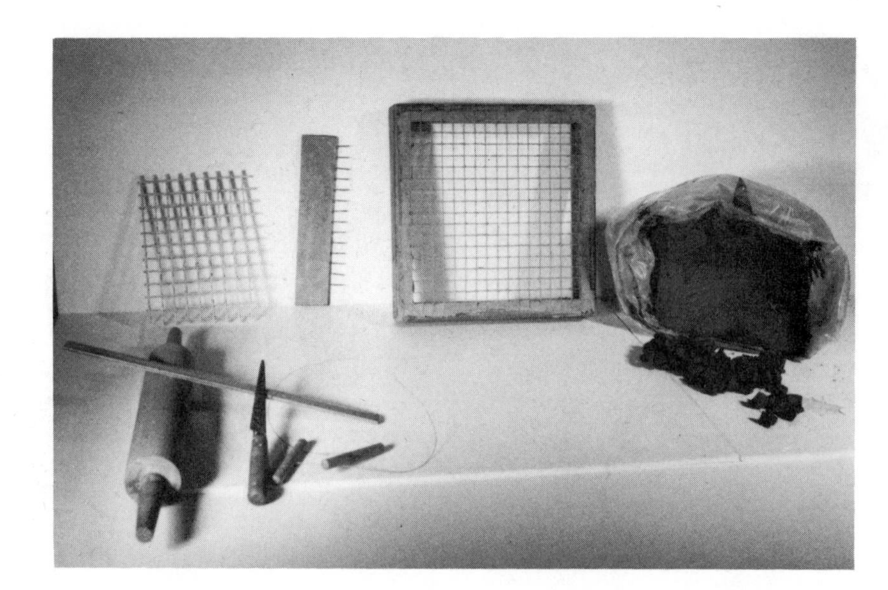

5.1 This picture shows two grids in addition to the tools and materials listed below. One is made with wires and a frame, and the other is cut from a plastic light diffuser. We can make small-scale adobes by rolling a slab and pressing it over the grid. The tool we choose to carry out the model building is, however, only the comb, standing next to the grid. The comb is the better tool.

TOOLS AND MATERIALS

The tools and materials we will use are very simple.

- A rolling pin used by potters or bakers
- Two bags of ready-made red clay for pottery or modelmaking, or a batch of clay mixed to a doughy consistency
- A piece of nylon or wire thread to cut the clay
- A long, sharp knife to cut the clay
- Two pieces of plywood 50 × 50 × 1 centimeters (¼ inch thick and 18 × 18 inches square). One piece will be used to prepare the material on, and the other is used for the model base.
- A piece of straightedge board 2 × 6 × 40 centimeters (1 × 3 × 18 inches)
- Two narrow pieces of wood 0.4 × 2 × 30 centimeters (3/16 × 1 × 12 inches)
- A comb made with a piece of board and thin 5-centimeter- (2–inch-) long nails. Clip the tops of the nails and drive them into the edge of a 5 × 15 × 1-centimeter- (2 × 6 × ¼-inch-) thick piece of wood. The nails must be 1.5 centimeters (½ inch) apart.
- A cup of water and a roll of paper towels or some old newspaper

HOW TO MAKE ADOBE BLOCKS, FLOOR SLABS, WALLS, ARCHES, DOMES, AND VAULTS

Adobe Blocks

Make a lot of little adobe blocks and save them before starting to build. To make blocks, take the following steps:

1. Cut and roll thin slabs 0.4 millimeters (3/16 inch) thick. The simplest way is to lay the two narrow pieces of wood on both sides of the clay block and pull the nylon thread over them to cut a thin slab. This eliminates the need to use a rolling pin.

5.2

5.3

2. Cut grooves with the comb into the thin slab. Push the comb all the way through so that the comb's teeth touch the plywood. Hold the comb at an angle for ease of work, and use the wood straightedge to make smooth lines. After practicing a while, freehand groove making will be easy.

3. Turn the board around and make grooves perpendicular to the previous ones to form the square blocks. This time, just press the comb part way into the slab to make strips of adobe blocks instead of individual blocks. The use of strips makes the work go faster.

Cut the block strips and save them under a plastic cover so that they stay moist. Also make some half-size blocks in the same way, by cutting the grooves half the distance of the full-size blocks.

Disregard other tools shown in the picture. They are alternative tools for block making.

5.4

Floor Slabs and Walls

5.5

1. To make a floor slab for a square room, roll a slab 1 centimeter (3/8 inch) thick, and cut a square from it measuring 20 × 20 centimeters (8 × 8 inches). Line the second piece of plywood with a piece of newspaper or paper towel, to keep the slab from sticking to the board. Place the slab on the lined board.

2. To practice making walls, mix some clay with water in a cup to make slip – a watery clay. This will be used for mortar. Use a finger to rub mortar all around the edges of the slab. Lay a two-adobe-wide wall over the mortar. Leave out four blocks for the door opening. As the walls rise, leave out four or more blocks for the window openings.

3. Begin to lay the second course of blocks over the mortar by placing one full adobe in the center and two half-adobes on the edges. Stagger the second layer's joints over the first course. With these two alternating courses, build up the wall all the way to the top, leaving openings for door and windows, as explained in Step 2. Use mortar throughout the work.

4. Build up the wall to about 10 centimeters (4 inches) high, and end the first course with two full-adobe courses. (Note that in these models solid clay walls are used instead of layers of adobe courses. After wall construction is learned, the model maker can use solid clay slab walls to save time and thus build more models.)

5.6

Arches

To make an arch, a form must be made first.

1. Cut a half-circle clay form as thick as the wall.

2. Locate the form over the edge of the window or door opening.

3. Put a piece of paper over the form for ease of removal.

4. Build the arch two blocks wide with mortar from both ends at once to join at the top. Tightly fit the last blocks to complete the arch.

5. Remove the form after the arch is completed and use it to build the other arch.

Domes

To make a squinch dome over the square room, four squinches must be built at four corners (for building a pendentive dome, see the chapter on domes in Part 3). To construct a squinch, follow these steps.

1. Use mortar on the top corner of two adjoining walls. Lay an adobe diagonally over the corner. Pitch two adobes on top of the first one, over mortar. Pitch three adobes over the existing pitched corner and continue to lay courses. Each course must slope a little toward the front, by using more mortar in the back of the courses. (See the life-size details in the squinch dome section in Part 3.)

5.7

5.8

5.9

5.10

2. When four squinches are constructed, we can go one of two ways:

One: Continue building layers over the four squinches until they meet each other at the base. Then continue with more layers and the same pitch in a herringbone pattern—adobes overlapping at the corners—to bring the dome to a square opening at the top. Fill the opening with more adobes until the dome is completed.

In life size this room is about 3 meters (10 feet) square. It has a door and window opening, and the interior dome structure will have a herringbone pattern. (It is similar to the squinch dome shown in the dome chapter.)

Two: When corner squinches are built, we can fill the gap between them with a single-adobe wall to the same elevation, so that a circular base is created for the dome. Then we build concentric circles until the dome is closed (see the chapter on domes).

5.11

The same steps explained above could be taken to build an octagonal dome. The only difference is that in this case we will have eight corners, which give us eight squinches instead of four. Thus to build a dome over a room with four, six, eight, or more walls, it is just a matter of constructing one squinch over each corner.

All door- and window-opening arches could be constructed at wall level, or projecting above the walls. The variations must be practiced with more models before beginning construction.

5.12

Vaults

1. Construct a rectangular room of 20 × 30 centimeters (8 × 12 inches), following the same steps as for the square room.

2. To construct a vault, an end wall (*espar*) must first be built, and then a leaning arch must be formed. A leaning arch is an arch built at an angle–instead of vertically–that leans on the *espar* or the previously built leaning arches. Thus it is necessary to build an espar first.

3. To construct an *espar*: Build a two-adobe-wide arch for the door and window. Lay adobes in their flat face over the arch to create a curved wall, which is the *espar*. (Note that the adobes in the models are strips instead of individual blocks to make the model making go more quickly.) The *espar* must be constructed to the desired height and curvature of the vault. (The espar *must* be two blocks wide to ensure strength. Figure 5.13 shows a one-block-wide espar, which, in reality, would not be practical.) It must also be wider than the room width, extending to the middle of the side walls where legs of the leaning arches end.

4. To construct a leaning arch: Lay one adobe, with mortar, against the *espar* on the inner edge of one of the side walls. This will be the first course. Lay the second course by starting with a half-adobe, and pitch with a full adobe against the *espar*. Lay the following courses so that a sloping leg of an arch is created from the base to the top of the *espar*. Build two half-arches this way from both sides. The seventh course must complete an arch that is one adobe thick at the top and seven adobes thick at the base. (Also see the illustration in the vault chapter in Part 3.)

5.13

5.14

5.15

5.16

5. Once the first leaning arch is formed against the *espar,* continue to lay adobes, with mortar, following the same curve as the first arch. Stagger the joints by alternating rows that start with a full- or half-size adobe at the base. In real life, the leaning arches are constructed from both sides and with the last block—the keystone adobe—at the top when the arch is completed. Use dry-packing (pieces of dry blocks used as wedges) between the adobe joints. For models the mortar may be enough. Thus, by succeeding courses of leaning arches, a vault is completed.

6. To create a skylight: Construct the vault from both ends of the room. Build a second *espar* on the opposite wall and start with a leaning arch. Construct rows of leaning arches from both ends until their bases meet where the skylight will be formed. Lay horizontal layers to fill the gap between the two vault sections. Leave a skylight opening at the top. Note: To accentuate the leaning arch geometry, a light colored clay is used in the models to construct some of the arches.

Pictures show a student's interpretation of a combination of a vault and a dome. Vaults could be constructed at one end with an *espar* and at the other end with dome squinches. In reality, once an arch, a dome, and a vault are constructed and their principles understood, unlimited combinations could be created.

5.17

5.18

5.19 Pictures show a combination of a vault and dome, a student's interpretation. Vaults could be constructed at one end with an espar and at the other end with dome squinches. In reality once an arch, a dome and a vault is constructed and their principles understood unlimited combinations could be created. Pictures show some of the models made by the students (of all ages).

5.20 Ojai Vault, an experimental vault constructed at the Ojai Foundation site, California, after a workshop. The material was dug from the site and the vault was built with small-size (8x8x2-inch) adobe blocks. The work was carried out under the direction of Mr. James Danisch, potter and ceramics professor. The 7x7x8-foot vault was the first adobe structure fired in the United States.

5.21 "Dome on the Range," a 16-foot-diameter, 13-foot-high parabolic dome built as an experimental structure in Bushland, Texas, by students after a workshop. The four arch openings on four sides of the dome are 5 feet wide at the base and 7 feet high—and also form a skylight. The dome was constructed with adobe blocks measuring 6x6x2 and 6x3x2 inches. The mortar is a mixture of clay and soda ash; it caused the blocks to fuse together in the firing process. The structure was fired through and, even though the blocks shrank approximately 8 percent, no cracks developed in the dome. The firing process took 17 hours and the temperature within the structure reached around 2100 degrees Fahrenheit.

The eight homemade burners used for firing were designed and manufactured based on the Mexican villagers' brick kiln burners using diesel fuel and steam; they were constructed from ¾-inch pipe pieces welded together.

The exterior of the structure was covered with fiber insulators (a fire blanket) to retain the heat and cause the structure to bake on the outside also from the reflected heat. In this case a high-tech material was utilized for a low-tech process, although mud-straw plaster over unbaked adobe blocks laid flat could have been used instead.

Project sponsors: Hunter Ingalls and Mary Emeny. Participants: Ellwood Pickering, Steve Haines, Alessandra Runyon, Marci Asner (students at Southern California Institute of Architecture); Mark Skinner, Bob Mekjian (students at California Polytechnic State University, San Luis Obispo); Hillary Guava of Texas; and Henry Oat, a potter and earth-builder from Santa Fe, New Mexico. (Photo by Marci Asner.)

CHAPTER 21
VISIONS FOR THE FUTURE

Building with earth and fire, air and water, with all or just one element in the spirit of unity, has created so many visions and dreams for me that it will take a separate volume to present them in a way that will do the Elements justice. Ultimately, the timeless materials and timeless principles of earth architecture must be comprehended and taken into the soul if they are to move in time to the forefront of the arts and sciences of the future. I will briefly outline some of those visions here, and I will end this book with a vision that moves earth architecture toward a new dimension: beyond our planet Earth.

- Earthen buildings can be used as kilns during their construction to produce both building materials and art pieces, which can in turn finance the cost of building. Thus low-cost housing can become no-cost housing.
- Many of the existing millions of earth structures in the world, especially those from China to Africa, built with suitable clay-earth, can be fired and glazed to become more durable, hygienic, and beautiful.
- Earth dams can be fired and glazed to become permanent.
- Sewage lines and underground water systems can be excavated and fired in place to create urban utility networks.
- Road and highways can be fired in place to form permanent surfaces.
- Buildings can be carved into the hills and fired, instead of being constructed over the ground. And ground can be carved and fired to become the buildings.
- Landscapes carved, fired, and glazed in place can be integrated into the contours of the site and architecture. Thus fire can become an important element of design in the landscape architecture of the future, as the element of water is today.

■ Ceramic sidewalks can be created in place to become part of a sculpted city landscape.

■ Manmade hills can be formed over city trash and piled in building layouts. The trash can be burned on the spot to strengthen the interior spaces, while the spaces themselves are being formed by the fire.

■ Underground oil, gas, and chemical tanks can be dug, fired and glazed to create permanent storage facilities.

■ Millions of bio-gas tanks needed for energy resources in the Third World that are now being built of concrete can be replaced by glazed and fired pits.

■ Troughs, irrigation channels, water reservoirs, and rain retention systems can be excavated and their voids used as kilns to produce building materials. Firing and glazing will cause their surfaces to become permanent.

■ Eroding coastal cliffs, mudslide hills, and unstable banks can be stabilized by the element of fire, in their natural settings, to form hard, brick-like surfaces and to fuse lava-layer rocks to slow the erosion process.

5.22 Eroded shoreline of the Pacific Ocean in Santa Monica, California.

■ Ice-reserve storage systems can be dug and fired underground to be used in the *Sardgah* system, for both cold-house storage and natural cooling of buildings, thus bringing both fire and ice to the service of humanity.
■ Sand dunes, which sometimes threaten to bury whole desert communities, can be stabilized by firing in place to form a glass layer at the top.
■ Volcanos pouring out millions of cubic meters of magma-lava materials can be utilized to generate materials for buildings and other spaces. Building complexes and integrated spaces can be planned and sand-earth casts prepared in anticipation ahead of an eruption to capture the molten earth flow. Sand-earth dug out from beneath cooled lava crusts can form organic, monolithic, and three-dimensional communities.
■ Fusion energy can be one of the answers to the world's housing shortage.
■ Earth and fire can become the therapeutic medium in the continuing cycle of human healing and rehabilitation.
■ Musical instruments formed from earth-clay can utilize fire and solar rays to create musical sounds, just as they are generated today by wind.
■ The elements of fire, water, air, and earth can be used to create inspired expressions of the arts, alone, or in the spirit of unity of the elements.

BEYOND OUR PLANET EARTH

As I was finishing the last pages of this book, I was invited with other scholars and scientists to present a paper about building on the moon to the NASA symposium on "Lunar Bases and Space Activities of the Twenty-First Century" held at the National Academy of Sciences in Washington, D.C. There I presented first the techniques of building with earth and fire on our planet, then the possibilities of constructing with the lunar soil, and the fire—the heat—of the sun, on the moon. The scientists' response was so overwhelming to me that what began as a vision has since changed to a search for design and construction possibilities.

At that presentation, the concepts of this book were briefly outlined to the audience: how we can build structures on our planet with earth or rock alone, and in harmony with gravity, by forming arches, vaults, and domes; how we can fire and fuse the structures built with earth to become like lava tubes (caves created by volcanic magma-lava, where the molten earth is its own form work); and how we should relearn the accumulated human knowledge of the timeless principles of earth architecture and ceramics and take them to the moon, Mars, and beyond. On the moon, which has one-sixth the gravity of Earth, our chances of building better and bigger are six times as high. And since the fire of the sun—solar heat—is abundant there, we can mold mounds of lunar soil to the desired building forms and melt these mounds with solar heat to create magma structures. The soil can then be dug out from under the hard crust, and packed on top for protection against high temperatures, radi-

5.23 Beyond our planet.

ation, and meteorite impact. In the low gravity of the moon, or in the zero gravity of space, we can build a giant potter's wheel and create ceramic structures. The space shuttle's covering of ceramic tiles points to the possibilities of the material. On the moon, we can also use focused sunlight to form fused adobe blocks to build our structure of arches, vaults, and domes. And finally, for the stabilization of the loose lunar dust, focused sunlight can be used instead of the chemical mixtures often proposed.

The full text of the paper is printed in the Appendix. While it may have a more complex tone than the simple language used in this book, the materials presented are but simple principles that can create unlimited possibilities in the future with the use and synthesis of the universal elements.

Earth, air, fire, and
water are obedient creatures,
they are dead to you and me,
but alive at God's presence.
—Rumi

Appendix

SHELL MEMBRANE THEORY APPLIED TO MASONRY DOMES

This paper was presented by professor Zareh B. Gregorian at the First Iranian Congress of Civil Engineering and Engineering Mechanics at Shiraz University, and later published in Art and Architecture, *Iran. (Original paper included photographs of the structures described in the text.)*

Shell membrane theory applied to masonry domes was known to builders. Construction was based on personal experience and the builder's knowledge, which was usually passed on from father to son and from master to apprentice.

During recent years, special attention has been given to the architectural form of domes and arches in Iran. The behavior of such forms, used to cover large spans, is of interest due to the fact that at the time of their construction, no other solution could be found. The lack of materials resisting tension and the consequent bending was overcome with the invention of steel, a revolutionary material which provided vast possibilities in the field of construction, leaving the masonry domes and arches forgotten.

Concrete is yet another solution to the creation of domes, due to its compressive strength and form acceptability. There is a very close similarity between the behavior of masonry domes and the behavior of concrete shell domes which are designed to follow the membrane theory. This may be seen from the stresses produced in masonry domes, which result in excessive and easily seen cracks from the form of construction at the base of the domes and from the fact that masonry domes with double curvatures cannot resist bending and can only withstand compressive stresses.

For comparison, it is necessary to refer to the membrane theory in shells. A very simple case of the membrane theory can be applied to spherical domes as follows:

Assuming a spherical dome with a radius of "r" and investigating the meridional force T and hoop force H, we can reach the following conclusions:

$$T = \frac{W}{2\pi r \, \text{Sin} 2\phi} \quad H = -T + rw \, \text{Cos}\phi_1$$

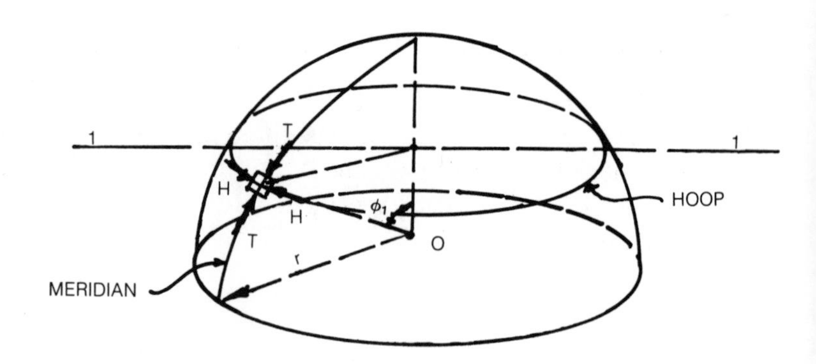

Or if we assume the weight of the dome acting on the top of plane 1-1 to be

$$W = 2\pi r^2 (1 - \text{Cos}\phi_1) \, w$$

which will yield T and H respectively:

$$T = \frac{2\pi r^2 (1 - \text{Cos}\phi_1) \, w}{2\pi r^2 \, \text{Sin}^2 \phi_1} = \frac{rw}{(1 + \text{Cos}\phi_1)}$$

$$H = \frac{-rw}{1 + \text{Cos}\phi_1} + rw \, \text{Cos}\phi_1 = Wr \left(\text{Cos}\phi_1 - \frac{1}{1 + \text{Cos}\phi_1}\right)$$

In order to investigate the variations of T and H at different points of the dome, it is enough to vary from zero to 90 degrees. The following will be the result:

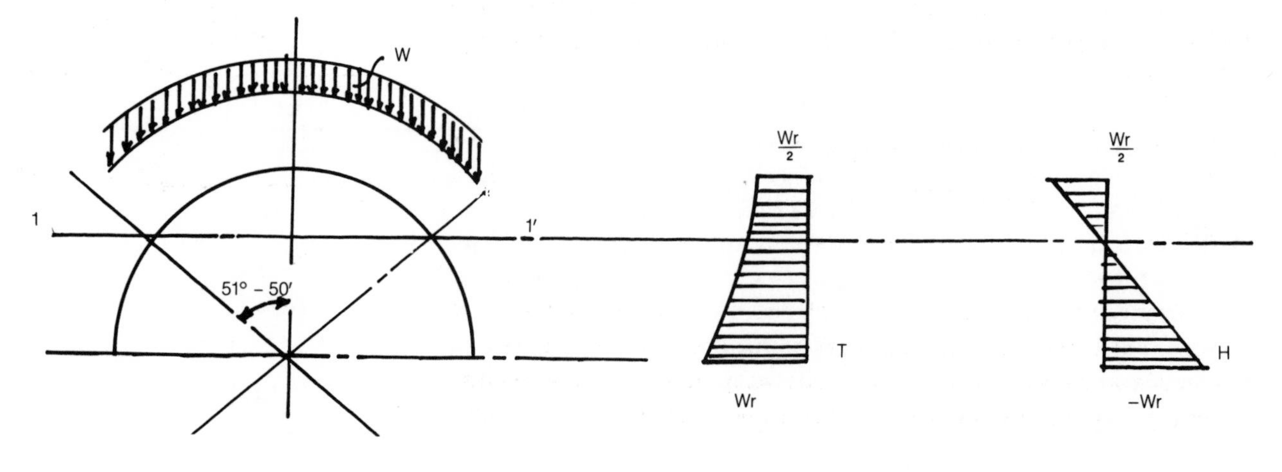

It is easily seen that T is always positive, that is, compressive as could have been predicted from the beginning. T increases from a minimum of $\frac{Wr}{2}$ at the top to a maximum of $\underline{Wr}$ at the base. The force H is compressive from the top of the dome to a plane 1′–1′ which has a central angle of 51°50′. From plane 1′–1′ to the base of the dome, the hoop force becomes tensile, increasing from zero to a maximum of $\underline{-wr}$ at the base. Thus, the following may be concluded:

A. Along the meridians there is always compression which increases with the angle ϕ_1

B. The force H acting on horizontal circles produces compression above the plane 1′–1′ and tension below plane 1′–1′. This tension increases from zero at plane 1′–1′ to a maximum at plane 0′–0′, causing cracks as shown on the diagram.

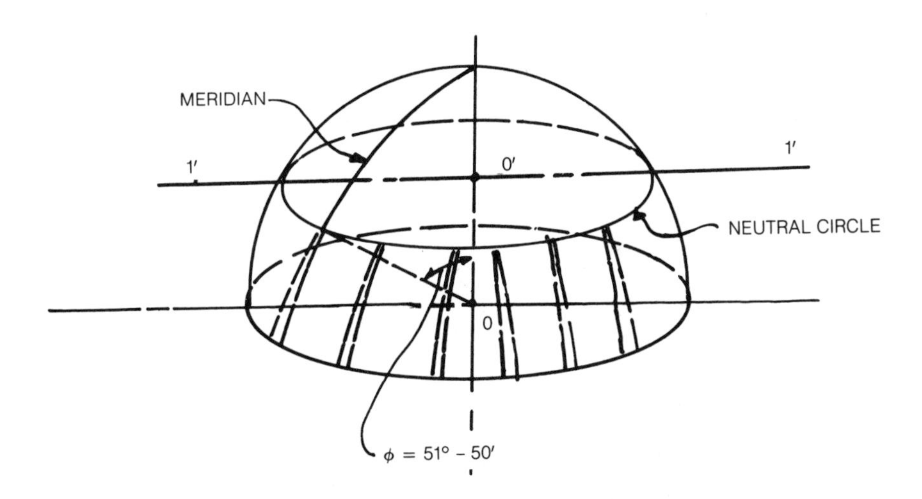

Investigation of the cross section of masonry domes shows that the thickness "t" increases from the top to the base in order to overcome the increasing compressive stress T and tensile stress H. Cracks seen on this type of dome result primarily from forces H and occur below the neutral circle along the meridians. This can easily be seen on the following existing buildings:

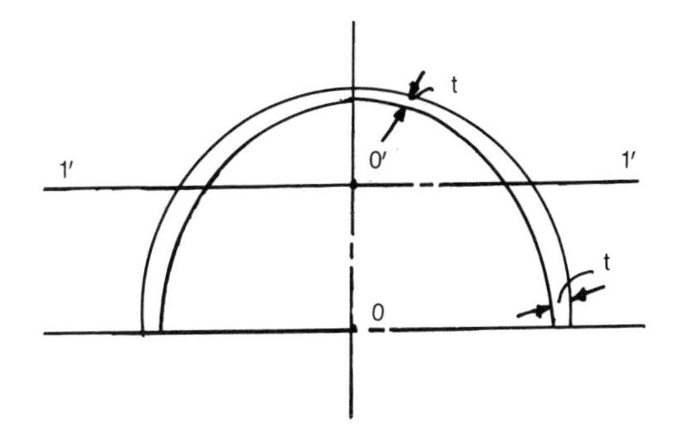

1. Dome of the Sultanieh Mosque, Zanjan

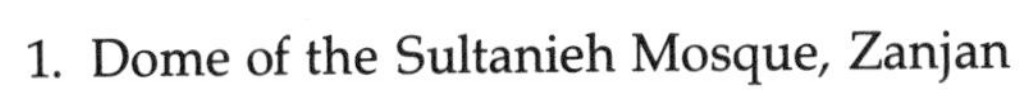

2. Armenian Vank Church, Djulfa-Isfahan
3. Mausoleum of Khwaja Rabi, Mashad

In order to resist the tensile force H, provisions have been made to increase the thickness of the dome where these forces exist or to reduce the tensile stresses by increasing the area through the use of heavy hexagonal or octagonal bondbeams around the periphery of the dome.

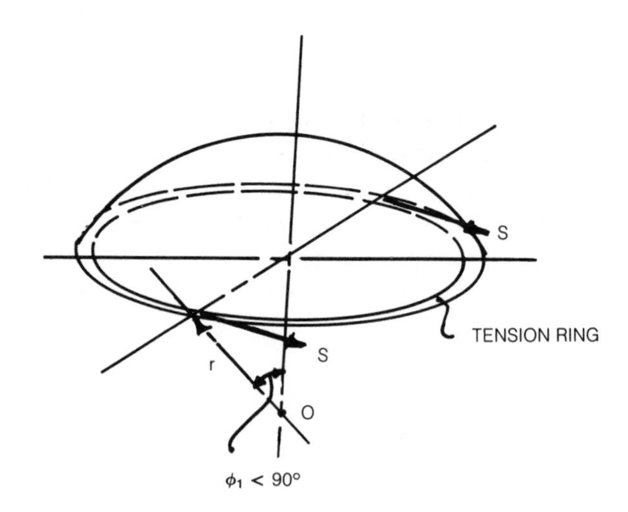

(Note: S is positive when $\phi_1 < 90°$, which indicates tensile force.) The tensile force becomes zero when $\phi_1 = 90°$.

$$S = \frac{rw \text{ Sin } 90° \text{ Cos } 90°}{(1 + \text{Cos } 90°)} = \frac{rw \times 1 \times o}{(1 + o)} = o$$

In several cases when the H stresses are calculated, small tensile stresses result which, though not acceptable theoretically, can be resisted in masonry domes by means of mechanical anchorage and bond between bricks, adhesion of mortar to bricks, or wooden bond beams. A special type of brick laying can be noticed in the lower parts of some domes. The resistance to tension brought about thereby should be investigated by means of small models put under loading.

Another significant point to be considered is the end circle of spherically incomplete domes. In such cases a tension ring must be provided around the periphery of the dome. The tension produced in such a ring as calculated by the membrane theory is equal to:

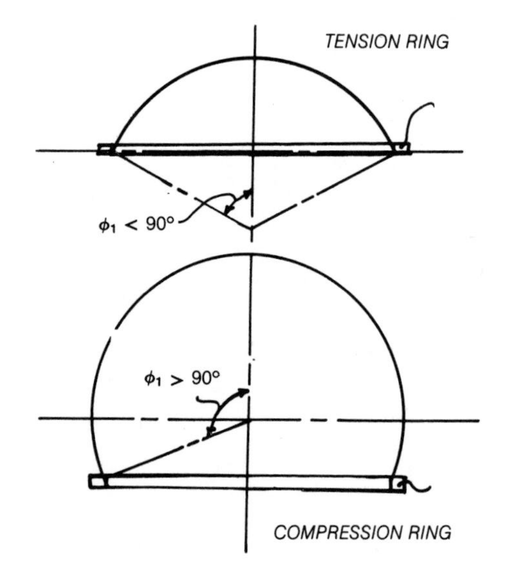

$$S = \text{Tr Sin}\phi_1 \text{ Cos}\phi_1 = \frac{rw \text{ Sin}\phi_1 \text{ Cos}\phi_1}{(1 + \text{Cos}\phi_1)}$$

For a complete, half spherical dome, theoretically no ring is required at the base when the dome is loaded uniformly under a load of "W" per unit of area.

Considering a spherical form larger than half a sphere, which will occur when $90° < \phi_1 < 180°$, we will get a negative value for S. This indicates compression in the ring. The following general conclusions may be derived.

C. When the base of a spherical dome occurs at an angle of $\phi_1 < 90°$, a tension ring is required around the periphery of the dome.

D. When the dome is a complete half sphere, no tension ring is required.

E. When the base of a spherical dome occurs at an angle of $\phi_1 > 90°$, a compression ring is required around the periphery of the dome.

This last form is the one most commonly used in masonry domes. The compression ring is obtained by a slight thickening of the dome at the base from the inside, providing a smooth, uniform surface on the exterior. Few domes have been made following the principle mentioned in item C, that is, with a central angle of $\phi_1 < 90°$. In such cases, a strong masonry ring is provided at the periphery of the dome to absorb the tension S, as can be seen in the dome of the Sheikh Safi in Ardabil which uses heavy hexagonal rings.

Domes which are half a sphere and employ no ring at the base are the most common type. Examples may be found in the Shrine of Mahan in Kerman , Masjid Sheikh Lotfollah in Isfahan and the Shrine of Sheikh Jabril in Ardabil.

Domes which are greater than half a sphere are also very frequent. A slight increase of the angle ϕ_1 from 90° provides the assurance of an existing compressive ring at the base. This is suitable for masonry construction and examples may be found in the dome of the Emamzadeh Ghassem in Tajrish and the dome of the tomb of Shah Chiragh in Shiraz.

There is also a basic comparison to be made between the shell action of conoidal domes and those of masonry. Analyzing the conoidal dome, we find:

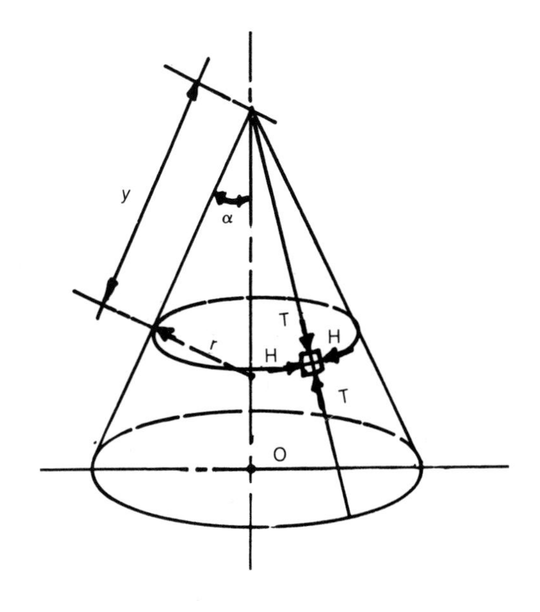

$$T = \frac{yw}{2\text{Cos}\alpha}$$

$$H = \text{wyty}\alpha\text{Sin}\alpha$$

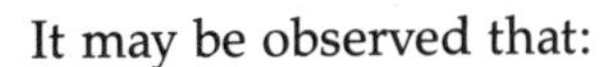

It may be observed that:

F. No tension problem exists as long as the angle $\alpha < 90°$. Thus, the form is very suitable for masonry work. Examples of this type are the dome of the Shrine of Ghabus Ibn Voshmgir, the dome of the Shrine of Alaeddin in Varamin and a tomb in Borujerd.

G. In some cases, domes are constructed with a combination of conoidal and spherical parts. In these cases, it may be assumed that the conoidal portions have no tension problems and the spherical sections are handled as described previously.

Final Conclusions

According to the membrane theory, no bending occurs in shells. Masonry domes also cannot accommodate bending because of the nontensile character of masonry work. Because of this, and due to the fact that the symmetry of a dome eliminates shear forces, the only existing forces are the meridional T and hoop H forces which have been discussed. This analysis is further strengthened by investigation of the visual and constructional characteristics of masonry domes, which relate to the basic principles of the membrane theory.

It is, therefore, appropriate that when we make use of the shell theory with modern materials such as concrete, acknowledging that what we do is a continuation of what our ancestors have done. By adapting what we have inherited to a more modern and scientific basis, we follow the traditional patterns in order to create new architectural forms.

Bibliography

1. Design of Circular Domes, Portland Cement Association
2. Thin Shell Structures, David Billington
3. Persian Architecture, Arthur Upham Pope
4. Persian Art, Andre Godard
5. Restoration Institute of National University of Iran

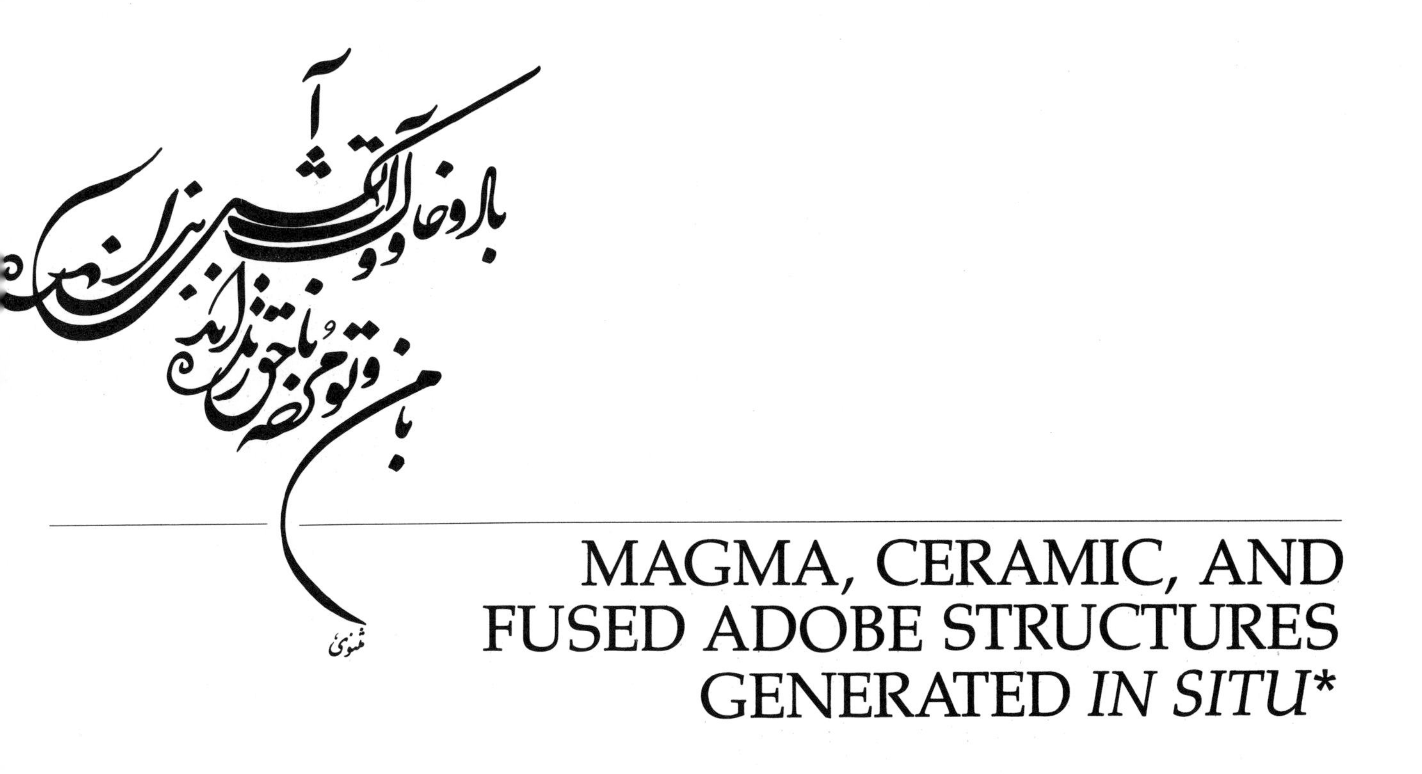

MAGMA, CERAMIC, AND FUSED ADOBE STRUCTURES GENERATED *IN SITU**

The accumulated human knowledge of "universal elements" can be integrated with space-age technology to serve human needs on Earth; its timeless materials and timeless principles can also help achieve humanity's quest beyond this planet. Two such areas of knowledge are in earth architecture and in ceramics, which could be the basis for a breakthrough—in scales, forms, and functions—in low gravity fields and anhydrous-vacuum conditions. With the added missing link of the element of fire (heat), traditional earthen forms can be generated on other celestial bodies, such as the Moon and Mars, in the form of magma structure, ceramic structure, and fused adobe structure. Ceramic modules can also be generated *in situ* in space by utilizing lunar or meteoritic resources.

TIMELESS MATERIALS–TIMELESS PRINCIPLES

The traditional techniques of building without centering, *i.e.*, leaning-arches, corbelling, and dry-packing can have greater applications in lower gravity fields, as well as higher material strength, than in the restricted conditions of these techniques' terrestial origins. At the same time, the "high-tech" heat-obtaining skills of solar heat, plasma, microwave, and melting penetrators can provide ceramic-earth shelters and appropriate technology for both developed and underdeveloped nations. Through understanding and utilizing the principles of "Yekta-i-Arkan"—unity of elements—integration of tradition and technology in harmony with the laws of nature is possible at many levels of microcosm and macrocosm.

*This paper was delivered at the symposium, "Lunar Bases and Space Activities of the 21st Century," organized by the NASA Johnson Space Center, in October, 1984 at the National Academy of Sciences in Washington, D.C.

MAGMA STRUCTURE

Lunar base structures can be generated and cast, based on the natural space formations created by magma-lava flow such as tubes and voids. By utilizing existing lunar contours or by forming mounds of lunar soil to desired interior spaces, structures can be cast *in situ* with the generated magma. Either way, the upper layers of the mounds and the apex, consisting of unprocessed lunar resources, can generate magma flow with focused sunlight (Criswell, 1976).

Ceramic-glass (Grodzka, 1976) and/or other lunar fluxes may be added to the main composite for lowering the melting temperature. Basalt melting point, 900° to 1200°C, can be lowered to glass composites' melting point with added lunar flux. As the molten composite flows with the low gravity crawl, the lava crust can be formed in spiral, circular, or multi-patterned rib troughs on the mound. A controlled flowing magma can cast single- or double-curvature monolithic shell structures. The underlying loose soil mound can then be excavated and packed over the monolithic shell for radiation/thermal/impact shielding (Carrier, 1976). Since high depth of necessary soil coverage over the structure is detrimental to both architectural flexibility and harmonious interaction of inner and outer space environments, the variable magma viscosity can be utilized to reduce the estimated 2-m thickness (Land, 1984) of the packed soil protections depending on material composites and attained temperature degree/time parameters.

The viscosity of the generated magma and the packed regolith can counterbalance internal atmospheric pressure, and the semi-glazed interior can provide an airtight membrane. The pliability of the magma medium can present new dimensions in the creation of sculptured interiors for the ultimate functional utilization of the generated spaces. It also offers an aesthetic dimension, since the molded forms conform to human generic non-angular tendencies. The organic material of magma and the possibilities for ceramic glazing of the interior will open a new era in integration of the arts to scales unattainable for humans under the limits of terrestrial conditions.

Magma materials, basaltic in particular, have produced agricultural soils and with suitable atmospheric conditions have proved to produce vegetation. Plant successions have taken place in magma-lava metamorphosis in terrestrial lava tubes and voids. Many examples of flora can be seen in old lava beds of the volcanic regions of the world. Similar conditions will be present in lunar magma structures when the temperature-moisture ambient exists for a life-supporting environment. Thus, common spaces of lunar bases could be designated as mini-agricultural zones that could both generate suitable atmosphere to sustain human life and provide supplemental nutrition resources.

Natural lava structures, such as Craters of the Moon National Monument, can provide case studies in the design development stages. Research is needed to determine material composites, magma crust formation patterns, and span limitations.

PREFABRICATED MAGMA MEMBERS

Conventional structures can be built with magma in lunar base complexes by prefabricating structural members. Beams, columns, panels, and connections can be prefabricated with generated magma composed of unprocessed lunar resources fused with solar heat. Magma-lava solidified structural members can be reinforced with fibers or reinforcing mesh produced from lunar resources. The precast panels and members can be post-tensioned by tendons, or fused with spot mortar composed of similar magma materials. Precast magma and ceramic members can be shaped to fit desired forms and functions. Lunar soil troughs and fused regolith layer form work can be utilized for casting systems.

CERAMIC STRUCTURE

The use of shielding ceramic tiles on the space shuttle points to the potential of ceramic materials for lunar and space applications. Ceramic structures of limited spans can be cast *in situ* on lunar sites; they can also be generated in space. On lunar sites, a centrifugally gyrating platform–a giant potter's wheel–featuring adjustable rims with high flanges can be utilized for the dynamic casting of ceramic and stoneware structures. A mass of lunar resources can be "thrown" in the stationary center zone of the platform and melted by focused sunlight to flow to the periphery rotating zone and cast desired shapes. Known lunar resources can also be spun on the same platform to create tensile fiber; by integrating the two operations, monolithic ceramic structures with tensile fiber reinforcing layers can be generated. Double-shell ceramic structures sandwiched with space and/or packed with insulating materials can provide radiation, thermal, and impact shielding. Such units can be used singularly for lunar camps or combined around a common hub and/or spine to form a lunar base complex.

The centrifugal platform system with its adjustable rim flanges can be utilized for lunar base infrastructure parts: pipes, ducts, and tunnel rings. Prefabricated sections for utility sheds can also be formed in single- or double-shell modules.

In space, a centrifugally gyrating platform moving in three dimensions can create more variations of ceramic structured modules than is possible in terrestial or gravity fields. Attached to a space station, the gyrating platform can generate ceramic modules *in situ*. The resources for ceramic structures can either be of lunar or martian origin or, in space, from captured meteoroids.

FUSED ADOBE STRUCTURE

Lunar base structures can be constructed *in situ* utilizing lunar adobe bricks produced from unprocessed lunar soil or the by-products of industrial mining operations. Lunar adobe blocks can be formed by the fusion of lunar resources with solar heat. It is anticipated that vacuum conditions and the essentially zero-moisture content of lunar soils should significantly reduce thermal diffusity (Rowley, 1984). Lunar adobe blocks can be used to build structures without form work, employing the earth-architecture techniques of dry-packing, corbelling, and leaning-arches (Khalili, 1986). The low gravity field and vacuum conditions, which allow for a smaller angle of repose and enhance lunar soil cohesion (Blacic, 1984), will give greater opportunity, in the case of the leaning-arch technique, for larger spans and shallower vaults and domes. The same advantages will cause the soil-packed covering to follow desirable contours for more flexible interaction of interior and exterior space and solar orientation. Fused spot-mortar or lunar dust sprayed at fusion point temperature can be used to bond the blocks in medium and large span structures. Arches, domes, vaults, and apses can be constructed to fit the contours of the moonscape; these curved surfaces can create sun and shade zones that are functionally desirable.

For functional or aesthetic reasons, total or partial interior ceramic glazing of lunar adobe structures can be done with lunar resources containing glass (Heiken, 1976) and other fluxes by solar heat fusion or plasma technology. The difficulty of mechanical separation of lunar dust can be solved by the bulk use of the soil at its powder stage, involving pre-heating the dust and igniting it on the structure at the point of fusion.

The techniques of earth-architecture and the human skills that have evolved to deal with natural materials and to meet the historic challenges of harsh environments and terrestrial gravity can put future men and women in direct touch with the lunar world. Discovering suitable dimensions of blocks, techniques of construction, and appropriate material composites while developing their own sense of unity with the lunar entity can be the start of human independence from Mother Earth, creating shelters in the heavens. The organic growth of lunar architecture, with its own materials and equilibrium of elements, can be used to initiate an indigenous and ecologically balanced human environment without damaging the heavenly body.

On Earth, one of the main tasks of architects, engineers, and builders has historically been nothing but winning the fight against gravity; now and in the future, the chance for victory on the Moon will be six times as great as it has been here on Earth.

INITIAL *IN SITU* CONSTRUCTION

Locating a lunar lava tube may well be one of the first stages of setting up a lunar base site. Lava tubes can provide the most expedient and economical way of starting an indigenous lunar architecture. Terrestial lava tubes are the best design model for exploring the development of appropriate life-supporting environments in lunar lava tubes. Either at the initial stage or in the following phases of lunar base construction, locating and utilizing lava tubes can be of great value.

An immediate construction system for the lunar base, after the initial camp setup, can utilize unprocessed lunar resources in a nonmechanized construction system. This system uses existing rocks of different sizes and dry-pack techniques. The low gravity field and higher rock fracture strength give added advantages for larger spans of corbelling and leaning-arch earth-structure systems. Meteroid and/or indigenous rock structures covered with lunar soil for radiation and thermal shielding can provide immediate, non-life-supporting shelters. Structures built with the same techniques can be fitted with an airtight fabric mesh for human habitation (Blacic,1984).

PAVING AND LUNAR DUST STABILIZATION

The lunar soil, with a particle size of about 70 microns, which adheres to everything and churns up with vehicular traffic, needs to be stabilized (Carrier and Mitchell, 1976). Fusion of the top layers of lunar soil with focused sunlight can form a magma-lava crust to arrest unstable lunar dust. Spacecraft landing pads, vehicular traffic roads, and pedestrian walkways can be paved with solar heat by on-spot fusion of the top layers, penetrating to desirable depth. Unprocessed lunar soil can be fused by solar energy via a manual or automatic control "paving" vehicle. Inappropriate regolith areas can be topped with a layer of appropriate lunar soil before its fusion. For low temperature fusion, lunar fluxes can be sprayed on top of the soil prior to introducing solar heat. Paving surfaces of heavier traffic areas can be constructed from composites fused to ceramic and stoneware consistency with desired colors and textures.

As a general rule, it is the use of the universal principles of the terrestrial element of fire (heat) – the solar rays – that must be thought of at the forefront of mediums and materials for planetary base design and construction. Adhering to the philosophy of the use of local resources, human skills, and solar energy, we can achieve our quests on the Moon, Mars, and beyond.

We must learn from the accumulated human knowledge of earth-architecture, which has sheltered humans in the harshest conditions. Each person going to the Moon, regardless of his or her work, must be aware of these fundamental principles and techniques to participate in creating an indigenous architecture to form their communities, not only because of economic benefit but also because of spiritual reward. As an old Persian saying goes, "Every man and woman is born a doctor and a builder – to heal and shelter himself."

ACKNOWLEDGMENTS

The Geltaftan Group, consisting of Manouchehr Sedehi, Mahamoud Hejazi, Ezzatollah Salmanzadeh, Ali Gourang, Ostad Asghar, and A. A. Khorramshahi, supported my work in earth-and-fire developments. Eyal Perchik, Alessandro Runyon, Tsosie Tsinhnahjinnie, Steven Haines, Ellwood Pickering II, Barclay Totten, Marci Asner, students at the Southern California Institute of Architecture, have helped advance my research work.

REFERENCES

Blacic J. D. (1984) Structural properties of lunar rock materials under anhydrous, hard vacuum conditions (abstract). In *Papers Presented to the Symposium on Lunar Bases and Space Activities of the 21st Century,* p. 76. NASA/Johnson Space Center, Houston.

Carrier W. D. III and Mitchell J. K. (1976) Geotechnical engineering on the Moon (abstract). In *Lunar Science VII, Special Session Abstracts,* pp. 92–95. Lunar Science Institute, Houston.

Criswell D. R. (editor) (1976) *Lunar Science VII, Special Session Abstracts (on Lunar Utilization),* pp. iii–vi. Lunar Science Institute, Houston.

Grodzka P. (1976) Processing lunar soil for structural materials (abstract). In *Lunar Science VII, Special Session Abstracts,* pp. 114–115. Lunar Science Institute, Houston.

Heiken G. (1976) The regolith as a source of materials (abstract). In *Lunar Science VII, Special Session Abstracts,* pp. 48–52. Lunar Science Institute, Houston.

Khalili N. (1986) *Ceramic Houses.* San Francisco: Harper & Row, 1986.

Land P. (1984) Lunar base design (abstract). In *Papers Presented to the Symposium on Lunar Bases and Space Activities of the 21st Century,* p. 102. NASA/Johnson Space Center, Houston.

Rowley J. C. (1984) In-situ rock melting applied to lunar base construction and for exploration drilling and coring on the moon (abstract). In *Papers Presented to the Symposium on Lunar Bases and Space Activities of the 21st Century,* p. 77. NASA/Johnson Space Center, Houston.

BUILDING CODES

NEW MEXICO ADOBE CODE, 1982

Section 2405. Unburned Clay Masonry

(a) **General.** Masonry of unburned clay units shall not be used in any building more than two (2) stories in height. The height of every wall of unburned clay units without lateral support shall not be more than ten (10) times the thickness of such walls. *Exterior* walls, which are laterally supported with those supports located no more than 24 feet apart, are allowed a minimum thickness of 10 inches for single story and a minimum thickness of 14 inches for the bottom story of a two story with the upper story allowed a minimum thickness of 10 inches. *Interior* bearing walls are allowed a minimum thickness of 8 inches. Upward progress of walls shall be in accordance with acceptable practices.

(b) **Soil.** The best way to determine the fitness of a soil is to make a sample brick and allow it to cure in the open, protected from moisture. It should dry without serious warping or cracking. A suitable adobe mixture of sand and clay shall contain not more than 2% of water soluble salts.

(c) **Classes of Earthen Construction**

(1) *Stabilized Adobes.* The term "stabilized" is defined to mean water resistant adobes made of soils to which certain admixtures are added in the manufacturing process in order to limit the adobe's water absorption. Exterior walls constructed of stabilized mortar and adobe require no additional protection. Stucco is not required. The test required is for a dried four-inch (4″) cube cut from a sample unit shall absorb not more than two and one-half percent moisture by weight when placed upon a constantly water-saturated porous surface for seven (7) days. An adobe unit which meets this specification shall be considered "stabilized."

(2) *Untreated Adobes.* Untreated adobes are adobes which do not meet the water absorption specifications. Use of untreated adobes is prohibited within 4 inches above the finished floor grade. Stabilized adobes and mortar may be used for the first 4 inches above finished floor grade. All untreated adobe shall have an approved protection of the exterior walls.

(3) *Hydraulically Pressed Units.* Sample units must be prepared from the specific soil source to be used and may be tested in accordance with approved test procedures.

(4) *Terrones.* The term terrone shall refer to cut sod bricks. Their use is permitted if units are dry and the wall design is in conformance with Sec. 2405 (a).

(5) *Burned Adobe.* The term "burned adobe" shall refer to mud adobe bricks which have been cured by low temperature kiln firing. This type of brick is not generally dense enough to be "frost proof" and may deteriorate rapidly with seasonal freeze-thaw cycles. Its use for exterior locations is discouraged in climate zones with daily freeze-thaw cycles.

(6) *Rammed Earth.*

1) *Soils:* See Section 2405 (b).

2) *Moisture Content:* Moisture content of rammed earth walls shall be suitable for proper compaction.

3) *Forms:* Suitable forms shall be used.

4) *Lifts and Compaction:* Uncompacted damp soil shall be compacted in lifts not to exceed 6" until suitable compressive strength is achieved.

5) *Tests:* Testing of rammed earth construction shall be in accordance with approved standards.

6) *Curing:* The building officials may allow continuous construction of rammed earth prior to the full curing process, provided proper compaction methods are followed.

(d) **Sampling.** Each of the tests prescribed in this section shall be applied to sample units selected at random at a ratio of 5 units/25,000 bricks to be used or at the discretion of the building official.

(e) **Moisture Content.** The moisture content of untreated units shall be not more than four percent by weight.

(f) **Absorption.** A dried four-inch (4") cube cut from a sample unit shall absorb not more than two and one-half percent moisture by weight when placed upon a constantly water saturated porous surface for seven (7) days. An adobe unit which meets this specification shall be considered "stabilized."

(g) **Shrinkage Cracks.** No units shall contain more than three shrinkage cracks, and no shrinkage crack shall exceed two inches (2″) in length or one-eighth inch (⅛″) width.

(h) **Compressive Strength.** The units shall have an average compression strength of 300 pounds per square inch when tested in accordance with ASTM C-67. One sample out of five may have a compressive strength of not less than 250 pounds per square inch.

(i) **Modulus of Rupture.** The unit shall average 50 pounds per square inch in modulus of rupture when tested according to the following procedures:

(1) A standard 4 × 10 × 14 cured unit shall be laid over (cylindrical) supports two inches (2″) from each end, and extending across the full width of the unit.

(2) A cylinder two inches (2″) in diameter shall be laid midway between and parallel to the supports.

(3) Load shall be applied to the cylinder at the rate of 500 pounds per minute until rupture occurs.

(4) The modulus of rupture is equal to $\frac{3WL}{2Bd^2}$

W = Load of rupture
L = Distance between supports
B = Width of brick
d = Thickness of brick

Footnote: Tests for pressed units is presently being developed.

(j) **Mortar.** The use of earth mortar is allowed if earth mortar material is of the same type as the adobe bricks. Conventional lime/sand/cement mortars of Types M, S, N are also allowed.

Mortar "bedding" joints shall be full SLUSH type, with partially open "head" joints allowable if surface is to be plastered. All joints shall be bonded (overlapped) a minimum of 4″.

(k) **Use.** No adobe shall be laid in the wall dependent on weather conditions until fully cured.

(l) **Foundations.** Adobes shall not be used for foundation or basement walls. All adobe walls, except as noted under Group M Buildings, shall have a continuous footing at least eight inches (8″) thick and not less than two inches (2″) wider on each side that support the foundation walls above. All foundation walls which support adobe units shall extend to an elevation not less than six inches (6″) above the finish grade.

Foundation walls shall be at least as thick as the exterior wall as specified in Section 2405 (l). Where perimeter insulation is used a variance is allowed for the stem wall width to be two inches (2″) smaller than the width of the adobe wall it supports. Alternative foundation systems shall be approved by the building official.

All bearing walls shall be topped with a continuous belt course of tie beam (except patio walls less than 6 feet high above stem). See "o" isolation piers.

(m) **Tie Beams.**

(1) *Concrete.* Shall be a minimum of six inches (6") thick by width of top wall. A bond beam centered to cover ⅔ of the width of the top of the wall by 6" thick shall be allowed for walls wider than 10". All concrete tie beams shall be reinforced with a minimum of two No. 4 reinforcing rods at each floor and ceiling plate line. All bond beam construction shall be in accordance with accepted engineering practices.

(2) *Wooden Tie Beam.* Shall be a minimum of 6" wall thickness except as provided for walls thicker than 10" above. Wood tie beams may be solid in the six inch (6") dimension or may be built up by applying layers of lumber. No layer shall be less than 1 inch (1").

(n) **Wood Lintels.** Shall be minimum in size six inches (6") by wall width. All ends shall have wall bearing of at least twelve inches (12"). All lintels, wood or concrete, in excess of nine feet (9') shall have specific approval of the building official. The building official shall approve all wooden tie beams for walls thicker than ten inches (10").

(o) **Anchorage.** Roof and floor structures will be suitably anchored to tie beams. Wood joists, vigas or beams shall be spiked to the wood tie beam with large nails or large screws.

Fireplaces shall be secured to the wall mass by suitable ladder reinforcement such as "durowall" or equivalent.

Partitions of wood shall be constructed as specified in Chapter 25, wood and metal partitions may be secured to nailing blocks laid up in the adobe wall or by other approved methods.

(p) **Plastering.** All *untreated* adobe shall have all exterior walls plastered on the outside with Portland cement plaster, minimum thickness ¾" in accordance with Chapter 47. Protective coatings other than plaster are allowed, provided such coating is equivalent to Portland cement plaster in protecting the untreated adobes against deterioration and/or loss of strength due to water. Metal wire mesh minimum 20 gauge by one inch (1") opening shall be securely attached to the exterior adobe wall surface by nails or staples with minimum penetration of one and one-half inches (1½"). Such mesh fasteners shall have a maximum spacing of sixteen inches (16") from each other. All exposed wood surfaces in adobe walls shall be treated with an approved wood preservative before the applicaiton of wire mesh. Alternative plastering systems shall be approved by the building official.

EXCEPTION: 1) Exterior patio, yard walls, etc. need not have Portland cement coating.

(q) **Floor Area.** Allowable floor area shall not exceed that specified under Occupancy. Adobe construction shall be allowed the same area as given in Type V-N construction.

(r) **Wall Insulation.** All methods of wall insulation shall comply with the manufacturer's recommendations.

(s) **Stop Work.** The building inspector shall have the authority to issue a stop work order if the provisions of this Section are not complied with.

Section 2405. Unburned Clay Masonry

UNIFORM BUILDING CODE, 1982

(a) **General.** Masonry of stabilized unburned clay units shall not be used in any building more than one story in height. The unsupported height of every wall of unburned clay units shall not be more than ten times the thickness of such walls. Bearing walls shall in no case be less than 16 inches. All footing walls which support masonry of unburned clay units shall extend to an elevation not less than 6 inches above the adjacent ground at all points.

(b) **Units.** At the time of laying, all units shall be clean and damp at the surface and shall have been stabilized with emulsified asphalt in accordance with U.B.C. Standard No. 24-14.

(c) **Laying.** All joints shall be solidly filled with Type M or S mortar. Bond shall be provided as specified for masonry of hollow units in Section 2410.

(d) **Stresses.** All masonry of unburned clay units shall be so constructed that the unit stresses do not exceed those set forth in Table No. 24-B. Bolt values shall not exceed those set forth in Table No. 24-C.

GLOSSARY

adobe A sun dried construction block made of earth-clay and sand; the essential building block of many earth structures.

adz A cutting tool with a curved blade that can be used to break adobes to size.

appropriate material Building material, usually the local earth or other natural material, that is well suited to local climatic conditions and building techniques.

appropriate technology Technology that is suited to the level of skills, resources, and construction techniques of a particular area, as opposed to the importation of advanced technology that requires specialized tools and skills.

arch A curved structure, kept in balance by the pull of gravity, which supports its own weight.

buttress A wall built at an angle for shoring and bracing; a counteracting force that strengthens and supports, as in a row of arches.

catenary A paraboloid curve; a catenary arch is in complete compression, strengthened by gravity.

cone 06 1,000°C (1,830°F); the average temperature for firing earth structures; the temperature at which a particular clay cone, cone 06, melts. (A cone is a small pyramid-shaped piece of material that indicates by bending or melting that a certain temperature has been reached.)

corbelling A building technique that involves stepping upward and outward from a vertical surface; corbelled domes, built of concentric rings of masonry blocks, are tall and conical.

dome A curved roof built over a square or circular room.

double dome A dome comprised of an inner shell and an outer shell with a layer of air sandwiched in between.

downdraft A type of kiln in which a fire is started below and the hot gases are exhausted through flues at the bottom.

dry-packing Dry pieces of broken bricks or rocks used to fill the empty space between adjoining adobes (fired structures must use only pieces of hard adobes or fired bricks).

earth architecture Structures made of earth, mud, or clay and based on a philosophy of harmony with nature.

earth-clay A mixture of earth and clay dug from the ground and used to make adobe blocks and mud-pile structures.

espar The curved end wall of a vault.

fire, firing The process of turning soft, wet clay into hard ceramic by baking it in a kiln.

flue An opening in a fired structure through which hot gases are exhausted during firing.

flux A material used in ceramics to create better fusion and lower the melting temperature of the mix. Colmanite (gerstley borate), soda, and powdered glass are common fluxes used in adobe structures.

form work (centering) A temporary structure of wood or metal that gives support to arches, domes, and vaults during construction.

geltaftan fired-structure (a compound of two Persian words: "gel" meaning clay and "taftan" meaning firing, baking, or weaving); the technique of firing earth structures to increase their durability (a *geltaftan* block may have a mixture similar to a fired brick, tile, or a volcanic rock).

glaze A clear or colored ceramic coating which, when fired, hardens to a waterproof and often glossy surface; a simple glaze can be made of powdered glass and oxides.

gravity-flow A firing system based on the flow of oil via gravity into a burner.

Hamedan burner A simple gravity-based kerosene oil burner used to fire earth structures.

kaval A short fired-clay pipe section used for lining the *qanats*.

kiln A room, small or large, where bricks, pottery, or other ceramic products are baked; in fired structures, the building itself becomes a kiln.

mortar An adobe mixture free from rocks and organic material, used between layers of adobe blocks to stick them together.

mud-pile (chineh) The least expensive and most quickly built earth walls; mud-pile walls are made by piling mud in a straight or tapered shape without form work, and allowing it to dry.

mud-straw A combination of mud and straw used to plaster adobe buildings; it improves stability, insulation, and waterproofing.

pendentive A triangular support that springs from the corner of a room to hold up the dome (commonly used in conjunction with arched walls).

qanat The main water irrigation system in Iran; a system of underground canals sometimes made up of short fired-clay pipe sections called *kavals.*

rammed earth A construction technique in which damp earth is rammed into a form to make a solid structure; it is especially good in damp climates.

skylight A window in the roof of a building designed to bring natural light inside the structure.

spring line The point on the wall from which an arch begins (springs).

squinch A diagonally built, cupped-hand-shaped interior or corner support that holds up a dome.

updraft A type of kiln in which a fire is started below and the hot gases are exhausted upward through a hole or chimney.

vault A single-curvature form that creates a ceiling in the form of a deep arch; it is built over long or rectangular rooms.

wind catcher A structure that brings cool outside air into a building; the most effective wind catchers work in conjunction with basements, pools, and fountains to create evaporative cooling systems.

METRIC AND U.S. SYSTEM OF WEIGHTS AND MEASURES

1 millimeter = 0.0393 inch

1 centimeter = 0.393 inch

1 meter = 39.3 inches or 3.28 feet

1 kilometer = 0.621 mile = 3,280 feet

1 square centimeter = 0.154 square inch

1 square meter = 1,549 square inches

1 cubic centimeter = 0.061 cubic inch

1 cubic meter = 35.31 cubic feet

1 gram = 0.035 ounce

1 kilogram = 2.204 pounds

1 metric ton = 2,204 pounds

1 liter = 1.056 liquid quarts

centigrade (Celsius) = °C

Fahrenheit = °F

°C = 5/9 (°F – 32)

INDEX